Intelligent Electronics and Circuits - Terahertz, ITS, and Beyond

Edited by Mingbo Niu

Published in London, United Kingdom

IntechOpen

Supporting open minds since 2005

Intelligent Electronics and Circuits - Terahertz, ITS, and Beyond
http://dx.doi.org/10.5772/intechopen.95654
Edited by Mingbo Niu

Contributors
Mingbo Niu, Xiaoqiong Huang, Hucheng Wang, Shangbo Wang, Ping Wang, Tongtong Shi, Rui He, Wubei Yuan, Biao Wang, Shaojun Lin, Sukhmander Singh, Shravan Kumar Meena, Ashish Tyagi, Sanjeev Kumar, Man Raj Meena, Sujit Kumar Saini, Salman Alfihed, Abdullah Alharbi, Krishnendu Raha, Kamla Prasan Ray

Notice
Statements and opinions expressed in the chapters are these of the individual contributors and not necessarily those of the editors or publisher. No responsibility is accepted for the accuracy of information contained in the published chapters. The publisher assumes no responsibility for any damage or injury to persons or property arising out of the use of any materials, instructions, methods or ideas contained in the book.

First published in London, United Kingdom, 2022 by IntechOpen
IntechOpen is the global imprint of INTECHOPEN LIMITED, registered in England and Wales, registration number: 11086078, 5 Princes Gate Court, London, SW7 2QJ, United Kingdom
Printed in Croatia

British Library Cataloguing-in-Publication Data
A catalogue record for this book is available from the British Library

Additional hard and PDF copies can be obtained from orders@intechopen.com

Intelligent Electronics and Circuits - Terahertz, ITS, and Beyond
Edited by Mingbo Niu
p. cm.
Print ISBN 978-1-80355-000-8
Online ISBN 978-1-80355-001-5
eBook (PDF) ISBN 978-1-80355-002-2

We are IntechOpen,
the world's leading publisher of
Open Access books
Built by scientists, for scientists

6,000+
Open access books available

146,000+
International authors and editors

185M+
Downloads

156
Countries delivered to

Our authors are among the

Top 1%
most cited scientists

12.2%
Contributors from top 500 universities

Interested in publishing with us?
Contact book.department@intechopen.com

Numbers displayed above are based on latest data collected.
For more information visit www.intechopen.com

Meet the editor

Dr. Niu received a BEng in Electronic Engineering from North-western Polytechnical University, China, and an MSc (Eng.) (first-class) in Communication and Information Systems from the same university. Prior to his Ph.D., Dr. Niu worked at a State Key Laboratory on underwater information and signal process-ing. He received his Ph.D. in Electrical and Computer Engineer-ing from the University of British Columbia, Canada. From 2008 to 2012, he was a research assistant at Optical Wireless Communications Laboratory and Integrated Optics Laboratory where he contributed to the development of ul-tra-high-speed optical data transmission links. Dr. Niu held a postdoctoral fellow-ship at Queen's University, Canada for two years. He also worked for Public Works at Calian Tech. Ltd., where he contributed to highly efficient statistical evaluation models of MIMO compressive sensing projects. Dr. Niu has co-authored more than thirty Institute of Electrical and Electronics Engineers (IEEE) and Optical Society of America (OSA) papers and supervised numerous student projects. Currently, he serves as a Lead Guest Editor for *Wireless Communications and Mobile Computing* and an Academic Editor for *IntechOpen*. He is a member of the Internet of Things (IoT) Committee at the China Institute of Communications (CIC). Dr. Niu received numerous scholarships during his undergraduate and graduate studies, includ-ing a Chinese Government Award, two University of British Columbia University Graduate Fellowships (UGFs), and a Huawei Tech. Ltd. Special Fellowship. His current research and teaching interests include the Internet of Vehicles (IoV), vehi-cle-to-road (V2R) infrastructure, cooperative microgrids, massive multiple input, multiple outputs (MIMO), image signal processing, low-carbon smart cities, energy harvesting, and electronic circuit theory. Dr. Niu is a licensed professional engineer in British Columbia.

Contents

Preface

Moving to smart grids to cyber-physical–social systems, moving from 5G to 6G, and moving from navigation assistance to intelligent driving, our world is changing and ubiquitous connections are beyond imagination. Emerging terahertz (THz) technologies, which could potentially change people's lives in every aspect, motivate us to merge, expand, and utilize THz bands for a variety of new designs, materials, and technique architectures by bundling integrated circuits, signal processing, energy harvesting, sensing, imaging, tomography, and data transmission.

With the increase of mobile traffic, several limitations in the existing social–physical world have emerged. Fortunately, new THz band and intelligent transportation system (ITS) technologies are promising candidates to combat these limitations. THz is an underdeveloped ultra-wideband zone between microwave and infrared that bridges the gap between electronics and optics. ITS can benefit from deep learning-based traffic smart electronics, control circuits, and vehicle-to-anything connections through road networks. To build extended wide-band systems that meet the aforementioned needs, researchers and engineers have introduced new advances in hybrid electro-photonics to THz technology and ITS fundamental designs. Great attention has been drawn to in-depth elaboration on information awareness and intelligent info-energy fusion.

THz has many applications across different research disciplines due to the advantages of electronics and optics. It is shaping a future wireless world and promoting initiatives on exploring brand-new integrated circuits for ITS, reconfigurable intelligent surfaces (IRS), nondestructive evaluation, simultaneous wireless information and power transfer, security, imaging, and signal processing.

This book provides a thoughtful and comprehensive understanding of the state of the art in THz science as well as novel technologies that promote essential aspects of contemporary ITS. It includes new research across different disciplines, from theoretical basis to experimental design, and from featured architectures to practical applications. The book is divided into four sections: "THz Science", "System Dynamics", "Smart City Essentials", and "Intelligent Device".

Section 1 discusses the science of THz technology. Chapter 1, "Broadband Terahertz Emission from Photoconductive Devices", discusses recent advances and studies on photoconductive components for THz emission, in which several materials have been employed to be photoconductive. It covers materials, quantum dots, nanostructure, dielectric materials, and the grating structure on photoconductive surfaces. It also examines screening effects. Chapter 2, "Studies of Terahertz Sources and Their Applications", focuses on THz radiation and its generation mechanism through laser-plasma interactions. It is pointed out that higher-power THz sources and their corresponding detector sets may be required for widespread applications.

High-frequency harmonics could cause serious system performance decay in intelligent automatic systems. If they are not smoothed, best performance cannot be achieved in an actual intelligent system. Section 2 includes Chapter 3, "Research and Application of PID Controller with Feedforward

Filtering Function", which studies industrial automatic system dynamics and proposes a new design for a PID controller with a filtering function, which offers weak noise amplification and strong harmonics' reduction with a smoothing effect.

Section 3 discusses advances in intelligent urban road networks that promote the development of smart cities. Chapter 4, "Traffic State Prediction and Traffic Control Strategy for Intelligent Transportation Systems", provides a comprehensive review of traffic status prediction and control techniques. Two deep learning approaches, convolutional neural network (CNN) and long short-term memory (LSTM) models are combined and realized for hybrid traffic state prediction with improved accuracy. Moreover, decentralized multi-agent advantage Actor-Critic technique and Nash Q learning is introduced for traffic signal control applications, being able to converge to the local optimum and overcome the scalability issue induced by city traffic policy updating for district neighborhood(s). Chapter 5, "Vehicle-To-Anything: The Trend of Internet of Vehicles in Future Smart Cities," discusses a variety of emerging technologies for intelligent roads and vehicles. Vehicle-to-anything (V2X) involves many different formats (V2V, V2R, V2P, V2I, and V2N) and all the elements on road with appropriate interconnection, including vehicles, infrastructure elements, people, information networks, and so on. Cooperative sensing, information query, intersection assistance, collision avoidance, and visible light positioning are introduced, which can be readily integrated into existing traffic infrastructure. For example, streetlights, direction signs, and traffic control signals can all be parts of V2X, often without the need for redeploying for enhancement of their basic functionalities. In dense urban areas, an increasing number of vehicles, especially private cars, raises a high demand for enhanced road capacity. Chapter 6, "Prediction of Large Scale Spatio-temporal Traffic Flow Data with New Graph Convolution Model", examines traffic flow as an important road condition, of which prompt and accurate prediction will provide better and more quickly adjusted traffic control and flow guidance. Part of smart city essentials is upgrading existing traffic control systems from "passive adjust" to "active control" and providing dynamic and flexible traffic control strategies. Thus, the chapter introduces a GCN-based traffic flow prediction model. Public traffic datasets worldwide are used for traffic prediction experiments, which include traffic data sources, data contents and data acquisition addresses, and a complete data processing process.

Intelligent devices make everyday life easier. Apart from mobile phones, mobility is an important factor to measure the convenience of an intelligent device. Section 4 discusses intelligent devices. Chapter 7, "Low-Cost Simple Compact and Portable Ground-Penetrating Radar Prototype for Detecting Improvised Explosion Devices", discusses the development and fabrication of a portable, compact, and low-cost continuous-wave (CW) ground penetrating radar (GPR) prototype operating at a single frequency of 920 MHz that applies enhanced isolation antennas. This device is demonstrated to be capable of detecting both metal and non-metal targets buried in soil as well as in a sandpit with high sensitivity. For example, it can detect a small bunch of wires buried 20 cm in the soil. Its maximum depth of detection in semi-dry soil is 65 cm for a metallic circular plate with a radius of 12.5 cm.

In summary, dedicated advances and the latest technologies in THz science, system dynamics, smart city essentials, and intelligent devices are presented in four sections of this volume. Though a fundamental understanding is provided for engineering and science communities on several selected topics, there exist many unmined work directions and research challenges. The chapters in this book can facilitate technology research and promote more excellent work in investigating the challenges in intelligent electronics and circuits fields for the development of a modern information society.

The editor would like to take this opportunity to thank all authors for their excellent contributions, the reviewers for their efforts in ensuring the high quality of the chapters, and the staff at In-Tech for their consistent effort and support.

Mingbo Niu
Chang'An University
Shaanxi, China

Section 1

THz Science

Chapter 1

Broadband Terahertz Emission from Photoconductive Devices

Salman Alfihed and Abdullah Alharbi

Abstract

This chapter explores the terahertz (THz) emission from biased semiconductor photoconductive devices. The photoconductive device is an optoelectronic device that is able to emit broadband THz radiation under the optical excitation, by an ultrafast laser, in the existence of a bias field. This chapter explains the basic principle of photoconductive devices with focusing on the main device components, being the photoconductive material and the photoconductive structure. Then, various materials and structures are discussed toward improving the performance of the photoconductive THz emitters. Furthermore, the main limitations and considerations are presented with insight into the different saturation and screening effects due to the bias field and pump fluence. Ultimately, the recent advances and studies of photoconductive THz emitters are presented in terms of material and structure, including the quantum dots, the nanostructure, the use of dielectric materials, and the grating structure on the photoconductive surfaces.

Keywords: photoconductive device, photoconductive THz emitter, semiconductor THz emitter, broadband THz emission, photoconductive antenna

1. Introduction

Since the ultrafast (femtosecond) laser was demonstrated in the 1980s, the field of terahertz (THz) technologies has emerged with an array of applications appearing in different areas, from spectroscopy and sensing to imaging and high-speed communications [1, 2]. Terahertz radiation is nonionizing radiation and has low photon energies, thus having less chance of tissues, cells, and DNA damage during the spectroscopic, sensing, and imaging applications. In addition, the terahertz radiation can be transmitted through some opaque objects in visible light, which opens an array of detection and security applications. The late development of the THz applications is due to the challenges in the generation and detection within the THz band. Its frequencies of 0.1 to 10 THz (30 µm to 3 mm), sandwiched between the electronic and optical frequencies, cannot be generated by conventional electronics or optical methods [3]. This is because the conventional electronics technologies are insufficient to produce broadband waves at these relatively high frequencies. On the other hand, conventional optical technologies cannot emit THz frequencies due to a fundamental issue; there is no material with a bandgap energy corresponding to the THz frequencies [4]. Fortunately, various ultrafast laser and semiconductors approaches have been examined and established. That leads to demonstration of the first emission of pulsed THz radiation using a dipole photoconductive antenna in 1988 by Smith *et al.* [5]. After that, the photoconductive devices were used widely to emit and detect broadband THz radiation; such devices have been developed with regards to their materials and structures to enhance the emission and detection of THz radiation. Nevertheless, the THz emission can be done mainly by two main methods, being optical rectification based on electro-optic (EO) crystal and photoconductive THz emitters based on semiconductors [6, 7].

The photoconductive THz emitter is an optoelectronic device with three main components, being the photoconductive materials, the photoconductive electrodes, and the lens [8] (**Figure 1**). The photoconductive material is a semiconductor having bandgap energy compatible with the photon energy of the ultrashort laser pulses. In addition, the photoconductive material should have optimum characteristics, including carrier lifetime, carrier mobility, breakdown voltage, and dark resistivity [9]. The carrier lifetime is preferable to be short. However, in the case of the photoconductive detector, it must be in the subpicosecond range. A higher breakdown voltage, carrier mobility, and dark resistivity are fundamental characteristics to assure a better photoconductive THz emitter performance in the form of higher radiated power, higher SNR, and broader bandwidth. The second

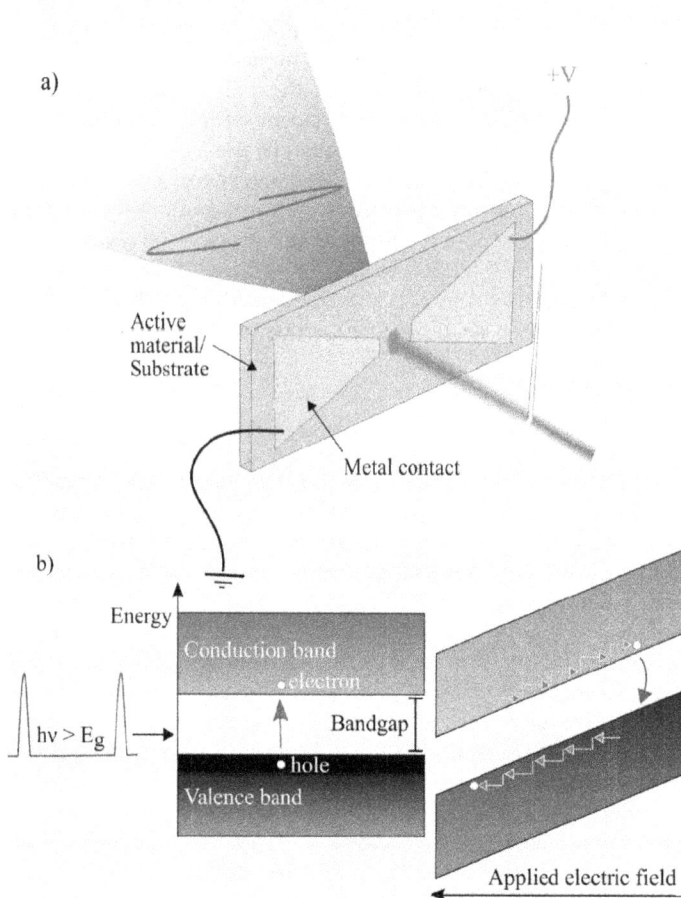

Figure 1.
Illustration of the photoconductive device, as in (a) it shows a schematic diagram of the photoconductive THz emitter, and in (b) it shows the semiconductor band structure under the applied electric field. © *IOP publishing. Reproduced with permission. All rights reserved [8].*

component is the photoconductive electrodes. The photoconductive electrodes are two metal electrodes patterned on top of the device, having a gap in between, namely, a photoconductive gap. The design and dimensions of such a gap will influence the device's performance. The last component is the lens. The lens is typically integrated with the emitter to accumulate the radiated field; the radiated field will then be focused on the targeted radiation path.

The photoconductive THz emitter can generate the THz radiation following photoexcitation of its photoconductive gap by an ultrashort laser pulse. When the laser pulse is focused into a photoconductive gap, the laser pulse generates free electrons and holes within the semiconductor, having a rate proportional to the laser pulse [10]. The free carriers will then accelerate under a bias field, controlled by the bias voltage, V, produce a transient photocurrent, and ultimately drive the emitter to emit far-field radiation with frequencies spanning into the THz spectrum [11].

This chapter presents the photoconductive devices for THz generation, with insights into their components, limitations, and considerations, and recent progress in this field. In Section 2, a number of photoconductive materials are discussed, the influence of the material and material's characteristics are addressed. In Section 3, the photoconductive electrodes (structure) are considered. This includes different structures characteristics based on their size, being a large aperture antenna, a small aperture antenna, and the plasmonic antenna, discussing the influences on the photoconductive THz emitters' performance. In Section 4, the limitations of the THz emission by photoconductive devices are discussed. The presented limitations are mainly related to the bias field and optical (pump) fluence, which appear in the form of radiated power saturation. In addition, the underlying physics of the space charge and near-field saturation is provided. Ultimately, in Section 5, the recent advances in photoconductive devices technology are given, including the integrated devices and the system-on-chip technologies.

2. Photoconductive materials

In general, the photoconductive THz emitter performance differs based on the photoconductive material and structure. Therefore, the photoconductive material will be the focus of this section. The optimum photoconductive materials would be crystal lattices with a direct bandgap between the valence and conduction bands. This bandgap determined the absorption wavelength of the exciting laser pulse. Other factors that play a significant role in choosing suitable materials are low carrier lifetime and high carrier mobilities. The most studied materials for photoconductive devices are gallium arsenide (GaAs), indium gallium arsenide (InGaAs), quantum well of InGaAs, indium aluminum arsenide (InAlAs), and a combination of group III-VI materials. This section will explore the photoconductive materials GaAs, ion-implantation in GaAs, InGaAs, and multi-quantum wells InGaAs/InAlAs.

Gallium Arsenide (GaAs) is a III–V semiconductor that has a bandgap of ($E_g \sim 1.42$ eV at 300 K) corresponding to the emission wavelength of 880 nm [12]. GaAs is compatible with the titanium-doped sapphire (Ti: sapphire) femtosecond pulsed laser sources typically used to illuminate the photoconductive THz emitters. The GaAs has been the most common material and is typically employed in semi-insulating (SI)-GaAs, low temperature-grown (LT)-GaAs, or ion-implanted GaAs. The SI-GaAs grown by liquid-encapsulated Czochralski at 450–600°C [13] is typically a single crystal that has a high resistivity ($>10^7$ Ω cm) and a high electron mobility ($\mu > 7000$ cm^2/Vs) [14]. The SI-GaAs is considered a cost-effective substrate and has become widely used material for photoconductive devices. However, the research was ongoing to shorten the carrier lifetime. The LT-GaAs grown on SI-GaAs is proved to reduce carrier lifetime two orders of magnitude to below 1 ps compared to SI-GaAs (t > 100 ps) and efficiently generate broadband THz radiations of over 1 THz with high resistivity (10^7 Ω cm) and reasonable mobility μ (100–300 cm^2/Vs) [15]. The growth is typically done by molecular beam epitaxy (MBE) on the surface of SI-GaAs substrate and growth temperature to between 200°C and 300°C in an arsenic-rich environment [16]. In such a case, it yields a high level of crystallinity, which means higher carrier mobilities and point defects due to excess As precipitants. Higher mobility leads to fast response, and point defects significantly reduce the lifetime (below 400 fs). These point defects act as recombination centers [15]. Increasing the temperature above 250°C will increase the lifetime to be greater than 50 ps. Tani *et al.* conducted a direct between LT-GaAs and SI-GaAs and studied the effect of growth temperature and anneal time effects where the carrier lifetime of LT-GaAs grown at 250°C and followed by post-growth annealing at 600°C for 5 min was found to be a 0.3 ps [17]. However, the process conditions for LT-GaAs are not easy to reproduce due to unreliable temperature monitoring below 400°C.

An alternative approach is using the ion-implantation technique to create point defects and reduce the lifetime in SI-GaAs by implementing arsenic, oxygen, nitrogen, carbon, and hydrogen (proton). Implanting H+ ions are shown to decrease the carrier lifetime of GaAs to sub-picosecond. Then several groups studied the effect of As^{+3} ion implantation of SI-GaAs and introduced excess As^{+3} impurities within the crystal structure similar to LT-GaAs [11]. However, the ion-implantation technique of As^{+3} (GaAs: As^{+3}) improved the controllability of the excess As^{+3} concentration and uniformity as compared to LT-GaAs, making it more reproducible than LT growth [11]. Salem *et al.* [18] characterized the GaAs:As$^+$ and unprocessed GaAs implanted with various other ions, including hydrogen, oxygen, and nitrogen. Their study revealed the lowest THz pulse intensity was observed in the GaAs:N^{3-} and all devices saturated at a higher pump fluence than nonprocessed GaAs emitter except GaAs:N$^+$. A study by Liu *et al.* [19] revealed that using multi-energy implanted As$^+$ ions leads to a shorter THz pulse and a higher bandwidth response than using single-energy ions.

The InGaAs are also employed as photoconductive material. It is a great advantage of the III-V compound to engineer the bandgap by changing the composition ratio. For example, the bandgap of the ternary compound indium gallium arsenide (In$_x$Ga$_{x-1}$As) can be potentially varied from 1.42 eV (x = 0) to 0.36 eV (x = 1). From a practical point of view, the protrontional to achieve 0.8 eV (1550 nm optical excitation) was the motivation for investigating this material for THz applications. Doping InGaAs by iron has been demonstrated to provide required recombination sites for a subpicosecond carrier lifetime, higher optical pump saturation power, and higher breakdown voltage. Wood *et al.* [20] investigated the InGaAs:Fe^{2+} emitter that is grown by using Metal organic chemical vapor deposition (MOCVD) across 830-nm to 1.55-μm optical excitation and found the highest THz power at the 1.2-μm excitation wavelength. This study shows precise control of the Fe-doping added during the epitaxial growth process and the strength of engineering the bandgap of III-V materials compound.

Heterostructure devices consisting of alternate InGaAs/InAlAs multilayer stacks (multiquantum wells) have been proposed [21] as potential materials for photoconductive devices and achieve high performance at 1550 nm comparable to LT-GaAs excited at 800 nm. Sartorius *et al.* demonstrated the first InGa(Al)As-based THz photoconductive devices operating at 1.5 μm [22]. In their device, an MQW comprised of 12 nm InGaAs:Be^{2+}/8 nm InAlAs as the photoconductive region was grown using standard low-temperature methods on an InP substrate. Moreover, the

Photoconductive material	Advantages	Disadvantages	Active layer	Operating wavelength (nm)
GaAS	The most used materials for THz PCAs and is well studied. It is the most efficient material for 800 nm.	It is not suitable for 1550 nm excitation wavelength.	LT-GaAs	780
			LT-GaAs	770
			LT-GaAs	776
			LT-GaAs	800
			SI-GaAs	800
			GaAs:Er	1550
InGaAs	Suitable for 1550 nm excitation wavelength.	Low dark resistivity.	InGaAs	1550
			InGaAs	1550
Multi-QW	Higher dark resistivity. High performance at 1550 nm comparable to LT-GaAs excited at 800 nm.	More complication.	InGaAs/InAlAs	1550
other materials of group III-V	The ability to engineer the target excitation wavelength.	More complication. It is not well studied.	InAs	780/ 1550
			InSb	780/ 1550
			GaSb	800
			GaAsSb	800 (up to 1440)
			InGaAs	800 and 1500
			GaInSb	800

Table 1.
Summary of some photoconductive materials with the advantages, disadvantages, active layer, and the operating wavelength.

material dark resistivity increases by four orders of magnitude comparable to bulk InGaAs due to the presence of Be^{2+} during the growth and the insertion of InAlAs, which has a higher dark resistivity than the InGaAs : Be^{2+}.

In addition to the GaAs, and InGa(Al)As, many other materials of group III-V such as InAs [23], InSb [23], GaSb [24], GaAsSb [25], and doped InGaAs [26], GaInSb [25] are studied as photoconductive material. Choosing the materials highly depends on the application and operating wavelength. Although LT-GaAs is still the most used material for photoconductive devices and is the most efficient material for 800 nm. However, it exhibits poor absorption at 1.55 μm, where other materials such as InGaAs or InGaAs/InAlAs heterostructure become more attractive. **Table 1** summarized some of the photoconductive materials with the advantages, disadvantages, active layer, and the operating wavelength.

3. Photoconductive structure

The photoconductive devices for THz emission have been developed extensively to fulfill the demand for high-performance THz emitters—and thus be essential for spectroscopic and imaging applications. The development of the emitters' structure is related to its design and dimensions and how that is attributed to the high performance of the THz emission. The performance of the photoconductive THz emitters is determined in the form of radiated power (or the THz spectral amplitude), SNR, and bandwidth. It is worth noting that the bandwidth here manifests itself as is the maximum frequency in the THz spectral amplitude, as a function of frequency, f, before the noise level of the measurements system. This chapter will provide an overview of such development regarding the design and structure of photoconductive THz emitters, with insights into the different emitter structures based on the dimensions, such as large-aperture and interdigitated electrodes THz emitters, small aperture THz emitters, and plasmonic THz emitters.

In the large-aperture and interdigitated electrodes photoconductive THz emitters, the gap between the two electrodes can be large as 4 mm to 130 μm [11]. Such a gap will allow a high level of optical excitation before reaching the saturation issues. Thus, the importance of such emitters stems from the need to scale up the radiated power, which is influenced by the incident optical power. A molded has been developed by Darrow *et al.* to study and predict the saturation in large-aperture photoconductive THz emitter, which is expected to have the saturation at higher pump fluence than the small-aperture THz emitter does [27].

In the small-aperture photoconductive THz emitter (dipole antenna), the gap between the two electrodes is smaller than in the large-aperture THz emitters, typically below 200 μm. In this case, it will be more difficult to align the laser spot within the PC gap. Although these emitters experience the saturation issues at lower pump fluence, in comparison with the large-aperture THz emitters, these emitters provide broader bandwidth over the large-aperture THz emitters. Our recent work on the design and structure of photoconductive THz emitters based on SI-GaAs examined the influence of bowtie structure characteristics on the THz spectral amplitude and bandwidth [28]. It is found that the bandwidth can be improved from 3.4 THz to 3.7 THz by changing the design of electrode structure from a sharp bowtie to an asymmetric bowtie structure at the same photoconductive gap. That could be attributed to the smaller capacitance of the sharp bowtie structure over the asymmetric bowtie structure, which results in a shorter resistance-capacitance (RC) time constant. The RC time constant, τ_{RC}, can be related to the gap conductance (of the photoconductive antenna), $G(t)$, the gap capacitance, C, and the antenna/transmission line impedance, Z_0 as [28]:

$$\tau_{RC} = 2Z_0 C / (2Z_0 CG(t) + 1), \tag{1}$$

Figure 2 illustrates the biased photoconductive gap with its equivalent circuit, here the redistribution of charge on the electrodes, can be seen as incident voltage waveform, $vi(t)$, reflected voltage waveform, $vr(t)$, and transmitted voltage waveforms, $vt(t)$.

The plasmonic THz emitter is introduced by Berry *et al.* by means of increasing the THz radiated power. An improvement of up to 50 times of the THz radiated power is observed using a plasmonic structure compared to a conventional photoconductive THz emitter [29]. The fabricated plasmonic THz emitter was based on LT-GaAs to provide an ultrafast photoconductor response. The distance between the two electrodes was 60 μm; the antenna structure was bowtie without fine tips; the maximum width of the antenna was 100 μm; the minimum width was 30 μm. In the plasmonic structure, the grating structure had a gap width of 100 nm, and the deposited gold had the same width of 100 nm. The optical pump focused on the gap close to the anode to maximize the THz radiation [29, 30]. **Figure 3** shows a comparison between the conventional photoconductive antenna and the plasmonic photoconductive antennae as in (a) the conventional photoconductive THz emitter and (b) the plasmonic

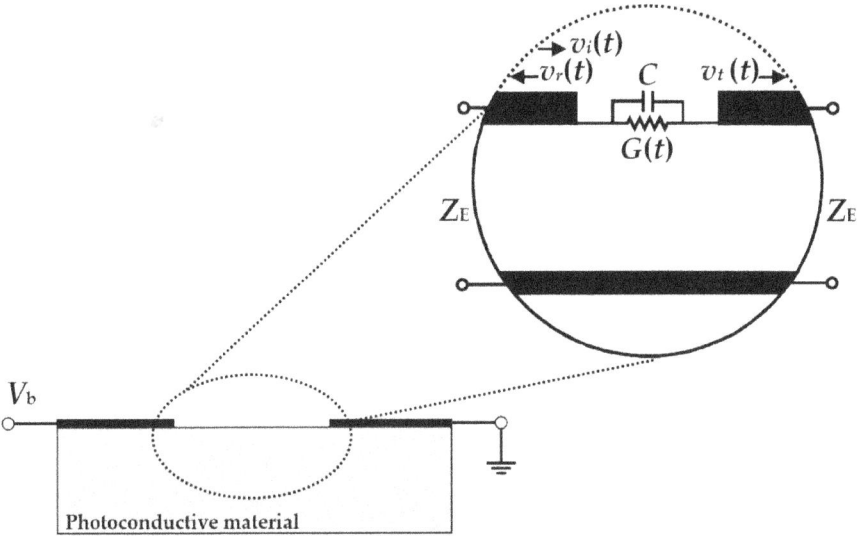

Figure 2.
A biased photoconductive gap at bias voltage, V_b. The inset shows an equivalent circuit of the photoinductive gap having the electrodes with incident voltage waveform, vi(t), reflected voltage waveform, vr(t), and transmitted voltage waveforms, vt(t). The gap conductance is shown as G(t) and the gap capacitance is shown as C. this figure is reprinted from [28].

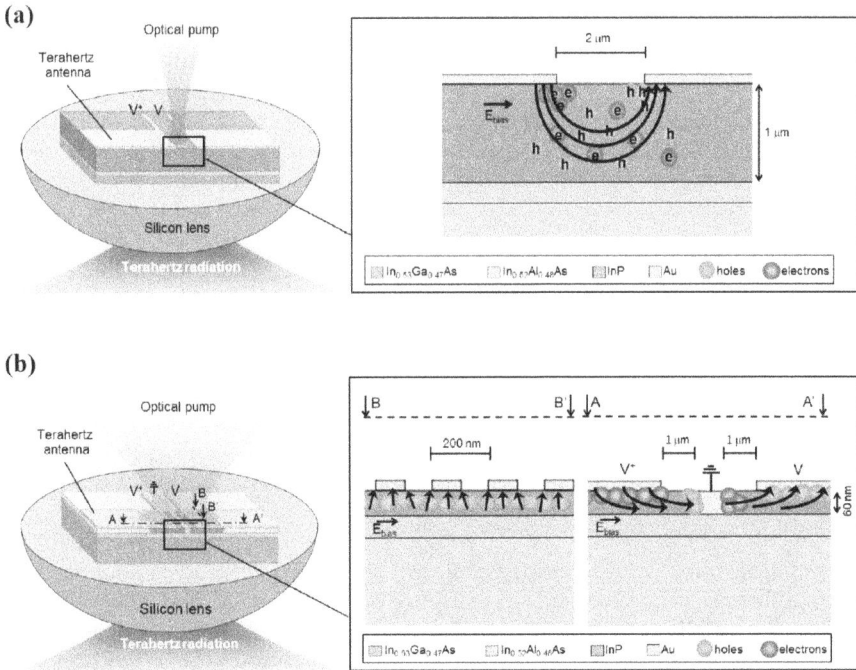

Figure 3.
A schematic diagram of the photoconductive device shows in (a), the conventional photoconductive THz emitter, and in (b), the plasmonic photoconductive THz emitter. © IOP publishing. Reproduced with permission. All rights reserved [31].

photoconductive THz emitter [32]. In addition to the plasmonic metal structure, the silver nano-islands on the photoconductive surface improve the performance of the photoconductive THz emitter [33]. Georgiou *et al.* have been recently demonstrated a 3-dimensional photonic-plasmonic photoconductive device. The performance of the THz emission is enhanced by employing a periodic array of nanopillars, which raise the optical absorption on the device surface and optimize the collection efficiency by converting each nanopillar into a single (nano)-photoconductive switch. As a result, nearly the overall generated current and the bandwidth are increased by 50-fold and five times, respectively. However, such a device will request high-tech technology to fabricate it [34]. On the contrary, the metal-based nano-islands structures enhance the photoconductive THz emission with a less complicated fabrication process. It is worth noting that these metal nano-islands can be produced by fine-tuning the deposition and thermal annealing process. Thus, it will offer less complexity during the fabrication in compression of the above-mentioned plasmonic structure.

Overall, the photoconductive structure plays a significant role in the performance of the photoconductive THz emitters. The large-aperture and interdigitated electrodes THz emitters mitigate the influence of saturation for scaling up the THz emission with the optical influence. The small aperture THz emitter (dipole antenna) shows a broader bandwidth, which allows discovering a more comprehensive range of THz frequencies. In addition, the recent studies on plasmonic devices present their significance to the photoconductive THz emitter performance. It also steers the future research and development of high-performance photoconductive devices for spectroscopy, sensing, and imaging applications.

4. Limitations of the THz emission by photoconductive devices

The underlying physics of the THz emission by photoconductive devices is the core of this section, which helps understand these devices' behavior. The photoconductive THz emission scales linearly with the applied bias field and pump fluence. However, that can be precise only in the ideal case, at low levels of bias field and/or optical excitation. Higher levels of bias field influence the photoconductive THz emitters' performance. Such influence can be seen as thermal effects, space-charge-limited current effects, etc. In addition, the photoconductive device has a limitation at a higher bias field correlated to the breakdown voltage of the photoconductive material. The pump fluence also has an impact, but that can be observed as the saturation of the THz radiation. The saturation (screening) of the THz radiation is mainly associated with two different mechanisms, being space-charge and near-field screening. This section will explore the limitation of the photoconductive THz emission with insights into the material and structure implications on photoconductive THz emitter's performance.

The THz radiated power (or the THz field amplitude, E_{THz}) can be scaled with the bias field, E_b, in three different mechanisms, based on the bias filed value and the photoconductive material, being a superliner (red), a linear (blue), and a sublinear (yellow), as shown in **Figure 4**. The superliner behavior is seen clearly with the photoconductive THz emitters based on GaAs. This superlinearity is associated with the space-charge-limited current due to the deep EL2 traps states [29, 35]. The superliner behavior is associated with the limitations within the semiconductor. This limitation manifests itself as Joule heating within the semiconductor and is observed in the photoconductive THz emitters based on InP [32]. Collier et al. have studied the Joule heating limitation in the photoconductive THz emitters based on InP. They found a correlation between the surface quality and the carrier lifetime, which directly affects Joule heating. In

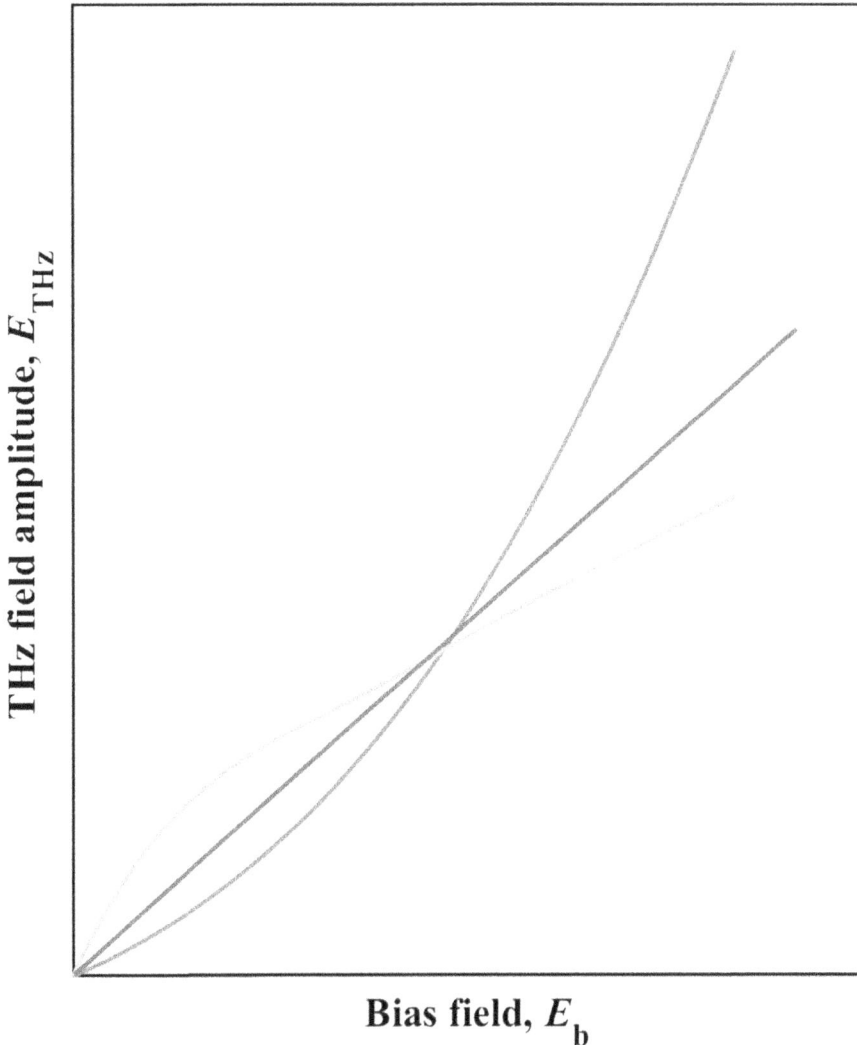

Figure 4.
The scaling of THz radiated power as the THz field amplitude, E_{THz}, with the bias field, E_b, in three different methods, being a superliner (red), a linear (blue), and a sublinear (yellow).

the textured InP photoconductive THz emitters, the carrier lifetime is decreased, which reduces the photocurrent and ultimately diminishes the Joule heating [32].

The pump fluence impacts the radiated THz power in the form of saturation (screening). At a higher level of optical excitation, the radiated THz power will be saturated. This saturation can be classified into two mechanisms, being space-charge and near-field screening. However, each screening status differs based on the photoconductive characteristics (material and structure) and optical characteristics (pump fluence). It is worth noting that transient mobility (mobility as a function of pump fluence) plays an important role in the emitted THz power and thus in the screening of the THz field [30]. The mechanisms of these two screening effects are explained in the next paragraph.

In the space-charge THz screening, the limitation of the photocurrent within the photoconductive gap is due to the high carrier densities within the photoconductive gap, affected by the high pump fluence. The charges drift in the opposite direction. Thus, the bias field screens and ultimately limits the radiated THz field [36]. In the near-field THz screening, the direction of the radiated THz field is in the opposite direction of the bias field, which limits the linear scale of the THz radiated field with the pump fluence, as increasing the pump fluence will raise the carrier densities within the semiconductor [37]. At the same pump fluence, the carrier densities in the emitter with a large photoconductive gap will be smaller than in the emitters with a small photoconductive gap. Thus, a large photoconductive gap emitter leads to scaling up the radiated THz power for higher levels, which increases the total emitter performance, before reaching the screening issues [30].

Overall, the main limitations of the THz emission by photoconductive devices can be related to the applied bias field and the exciting pump fluence. The two limitations are correlated with the photoconductive material and structure characteristics. These two limitations prevent the THz field amplitude from scaling linearly with the bias field and pump fluence. Thus, it is essential to design the photoconductive THz emitter carefully. Furthermore, the photoconductive material must be chosen judiciously to meet the demand of the high-radiated THz field for the aforementioned advanced applications.

5. Recent advances in photoconductive devices

A number of the recent advances and research in the field of photoconductive devices are discussed in this section, with insight on the development of the material and structure to enhance the photoconductive THz emission for spectroscopic, sensing, and imaging applications. The section will explore different approaches including:

- Quantum dots.

- Nanostructured electrodes (non-plasmonic) of the photoconductive device.

- Dielectric metasurfaces in photoconductive terahertz devices.

- Grating photoconductive devices.

The development of the photoconductive THz emission using such new approaches is notable. The quantum dots are mainly related to photoconductive materials. In contrast, the nanostructured electrodes, dielectric metasurfaces, and Grating photoconductive devices are associated with the photoconductive structure. Here, the main interest is to focus on improving the THz emission using these different approaches and the potential enhancement of these devices.

The quantum dots have been employed to boost the photoconductive THz emitters' performance. Gorodetsky *et al.* used InAs quantum dots in bulk GaAs. The short carrier lifetime has captured with the dots in such devices and maintains high carrier mobility [38, 39]. The photo-electronic priorities of the quantum dots can be managed by controlling the characteristics of these dots during epitaxial growth. It is worth noting that quantum dots have three-dimensional effects compared with a one-dimensional effect in the quantum wells. In Gorodetsky *et al.* study, the active regain consists of InAs quantum dots layer (1–2 nm), InGaAs wetting (5 nm), and GaAs spacer (35 nm). At the top of these layers, an LT-GaAs layer is grown with a 30-nm thickness, the observed boost of such structure is about 5-fold at 1.0 THz [37]. In addition, GaAs with ErAs quantum dots has been demonstrated for exaction laser having a 1550-nm wavelength [40]. The observed conversion rate (from optical to THz power) is 0.18%. However, this result was obtained by using a resonant slot antenna.

Nanostructure electrodes of the photoconductive device show an improvement of the photoconductive THz generation, even without a plasmonic effect. Although the plasmonic photoconductive THz emitter is one of the breakings through in the THz generation and detection field, the nanostructure has its encasement on the performance of such devices [41]. Singh et al. examined an antenna nanostructure fabricated by utilizing an electron-beam lithography system (EBL), having a 5-nm titanium layer and a 25-nm gold layer. Hilbert-fractal design is used with different line widths up to 140 nm. An improvement of the emitted THz power by an approximate factor of two is observed using this nanostructure.

Dielectric metasurfaces in photoconductive terahertz devices can be used as an alternative method to enhance the photoconductive THz emitters' performance instead of the plasmonic structure [42]. Although the plasmonic structure delivers better THz field improvement over the dielectric structure, the dielectric structure has a substantial characteristic which is the lack of dissipation [43]. In addition, the optical absorption of the incident light (laser) onto the photoconductive device can be improved by reducing the Fresnel losses, which can be done by using thin films of dielectric materials on top of the photoconductive gap. These dielectric materials (thin films) include SiO_2, Si_3N_4, Al_2O_3, and TiO_2 [44, 45]. **Figure 5** shows a bowtie antenna having a layer of TiO_2 being coated on the photoconductive gap, in (a) the schematic view of the photoconductive THz emitter, (b) the SEM image of the photoconductive THz emitter, and (c) the THz spectral amplitude obtained with using TiO_2 layer (red) and without using TiO_2 layer (black), "from [45]".

The grating structure manifests itself as a periodic array of grooves, lines, slits, etc. The grating structure of the photoconductive devices for THz generation has been studied according to the effective medium approximations (or effective medium theory). The theory can be applied to describe the interaction of light with the grating structure (subwavelength) [46]. Chia et al. have modeled and simulated the influence of grating structure on the THz emission performance by COMSOL Multiphysics software with an insight into the effects of grating geometrical parameters. The author funds an improvement of about 1.63 of the photocurrents obtained by an optimized grating structure of photoconductive THz emitter over the planer emitter structure. This is due to the higher photon absorption, which leads to and leads to more carrier generation within photoconductive material, thus higher photocurrent is observed [46]. **Figure 6** shows the simulated grating structure of LT-GaAs and its effects, as in a) the upper diagram shows the surface of

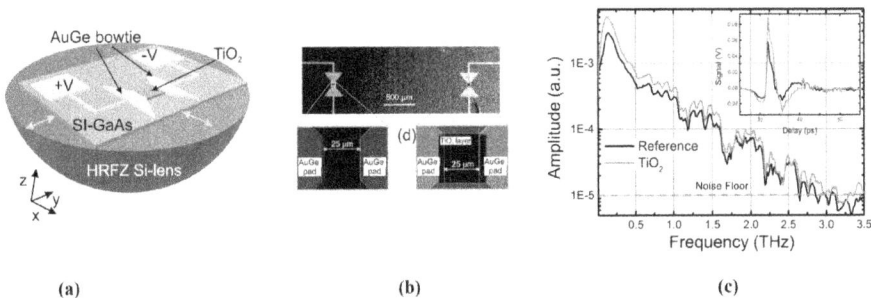

(a) (b) (c)

Figure 5.
The bowtie photoconductive antenna with TiO_2 layer, coated on the photoconductive gap, in (a) the schematic view of the photoconductive THz emitter, (b) the SEM image of the photoconductive THz emitter, and (c) the THz spectral amplitude obtained with using TiO_2 layer (red) and without using TiO_2 layer (black). This figure is reprinted from [45] with the permission of AIP publishing.

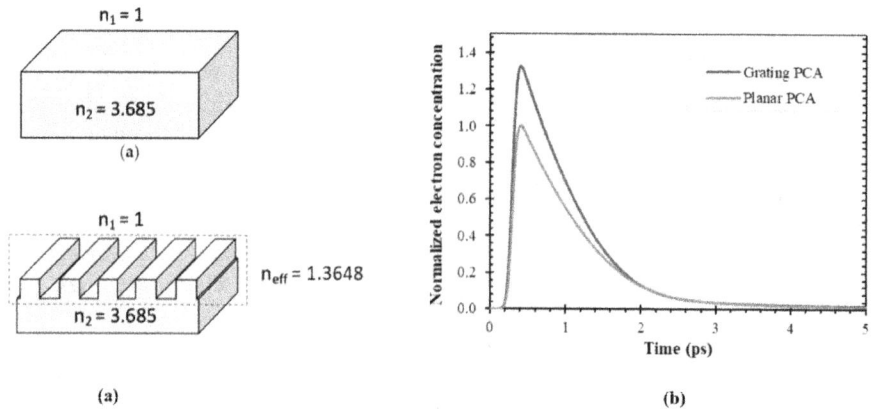

Figure 6.
The simulated grating structure of LT-GaAs, as in (a) the upper diagram shows the surface of planner photoconductive THz emitter, the lower diagram shows grating structure of the photoconductive THz emitter, and (b) the normalized electronic concertation obtained by the two different simulated photoconductive THz emitters. This figure is reprinted from [46].

planner photoconductive THz emitter, the lower diagram shows grating structure of the photoconductive THz emitter used in the simulation, and b) the normalized electronic concertation obtained by the two different simulated photoconductive THz emitters, "from [46]".

Nowadays, the development of photoconductive devices regarding materials and structure is a hot research topic. Several publications have discussed many schemes to achieve higher performance of THz generation by photoconductive devices to facilitate the applications in cutting-edge technologies such as THz spectroscopy, THz sensing, and THz imaging. For photoconductive materials, the research focuses on the quantum dots as well as promotes material properties such as the carrier lifetime and carrier mobility. For the photoconductive structure, the implementation of plasmonic and nanostructures shows its advantage for the aforementioned applications. However, utilizing some novel ideas such as grating structure and a precise selection of the dielectric material is demonstrated to boost the performance of photoconductive devices further.

6. Conclusion

This chapter presented the photoconductive devices for THz emission. Several materials have been employed as photoconductive materials. However, GaAs is a typical material for these applications, particularly for the sapphire femtosecond pulsed laser sources, which emit at the same range of the bandgap energy of GaAs. Furthermore, several photoconductive structures have been employed. The plasmonic structure shows the highest impact of the photoconductive THz emitters' performance over the microstructure photoconductive THz emitters. On top of that, the screening effects of the THz field amplitude is an issue limiting the linear scaling of the THz field with the pump fluence. Such limitations can be diminished using a large-aperture photoconductive antenna. At the end of this chapter, the improvement of these devices' performance has been considered by viewing some recent work in this area. The work has also presented the influence of the quantum dots, the nanostructured electrodes (nonplasmonic) of the photoconductive device, the dielectric materials in photoconductive terahertz devices, and the grating structure on the photoconductive surface. It is hoped that the presented work can lay a role in continuing advancements of photoconductive devices.

Conflict of interest

The authors declare no conflict of interest.

Author details

Salman Alfihed* and Abdullah Alharbi
King Abdulaziz City for Science and Technology, Riyadh, Saudi Arabia

*Address all correspondence to: salfihed@kacst.edu.sa

IntechOpen

References

[1] Nagatsuma T, Ducournau G, Renaud CC. Advances in terahertz communications accelerated by photonics. Nature Photonics. 2016;**10**:371-379

[2] Walther M, Fischer BM, Ortner A, Bitzer A, Thoman A, Helm H. Chemical sensing and imaging with pulsed terahertz radiation. Analytical and Bioanalytical Chemistry. 2010;**397**(3): 1009-1017

[3] Kleiner R. Filling the terahertz gap. Science. 2007;**318**:1254-1255

[4] Lewis RA. A review of terahertz sources. Journal of Physics D: Applied Physics. 2014;**47**:374001

[5] Smith PR, Auston DH, Nuss MC. Subpicosecond photoconducting dipole antennas. IEEE Journal of Quantum Electronics. 1988;**24**(2):255-260

[6] Alfihed S, Holzman JF, Foulds IG. Developments in the integration and application of terahertz spectroscopy with microfluidics. Biosensors and Bioelectronics. 2020;**165**:112393

[7] Rice A, Jin Y, Ma XF, Zhang X-C. Terahertz optical rectification from⟨110⟩ zinc-blende crystals. Applied Physics Letters. 1998;**64**(11):1324-1326

[8] Bacon DR, Madéo J, Dan KM. Photoconductive emitters for pulsed terahertz generation. Journal of Optics. 2021;**23**(6):064001

[9] Ferguson B, Zhang X-C. Materials for terahertz science and technology. Nature Materials. 2002;**1**:26-33

[10] Uhd Jepsen P, Jacobsen RH, Keiding SR. Generation and detection of terahertz pulses from biased semiconductor antennas. Journal of the Optical Society of America B. 1996;**13**(11):2424-2436

[11] Burford NM, El-Shenawee MO. Review of terahertz photoconductive antenna technology. Optical Engineering. 2017;**56**(1):010901

[12] Venkateshm M, Rao KS, Abhilash TS, Tewari SP, Chaudhary AK. Optical characterization of GaAs photoconductive antennas for efficient generation and detection of terahertz radiation. Optical Materials. 2014;**36**(3):596-601

[13] Murotani T, Shimanoe T, Mitsui S. Growth temperature dependence in molecular beam epitaxy of gallium arsenide. Journal of Crystal Growth. 1978;**45**:302-308

[14] Clegg JB, Makram-Ebeid S, Tuck B. Semi-Insulating III-V Materials. United State: Evian; 1982. pp. 80-91

[15] Gupta S, Frankel MY, Valdmanis JA, Whitaker JF, Mourou GA. Subpicosecond carrier lifetime in GaAs grown by molecular beam epitaxy at low temperatures. Applied Physics Letters. 1991;**59**(25):1.105729

[16] Jooshesh A, Bahrami-Yekta V, Zhang J, Tiedje T, Darcie TE, Gordon R. Plasmon-enhanced below bandgap photoconductive terahertz generation and detection. Nano Letters. 2015;**15**(12):8306-8310

[17] Tani M, Matsuura S, Sakai K, Nakashima S-i. Emission characteristics of photoconductive antennas based on low-temperature-grown GaAs and semi-insulating GaAs. Applied Optics. 1997;**36**(30):7853-7859

[18] Salem B, Morris D, Aimez V, Beerens J, Beauvais J, Houde D. Pulsed photoconductive antenna terahertz sources made on ion-implanted GaAs substrates. Journal of Physics: Condensed Matter. 2005;**17**(46): 7327

[19] Liu T-A. THz radiation emission properties of multienergy arsenic-ion-implanted GaAs and semi-insulating GaAs based photoconductive antennas. Journal of Applied Physics. 2003;**93**(5): 1.1541105

[20] Wood CD, Hatem O, Cunningham JE, Linfield EH, Davies AG, Cannard PJ, et al. Terahertz emission from metal-organic chemical vapor deposition grown Fe:InGaAs using 830 nm to 1.55 μm excitation. Applied Physics Letters. 2010;**96**:194104

[21] Dietz RJB, Globisch B, Gerhard M, Velauthapillai A, Stanze D, Roehle H, et al. 64 μW pulsed terahertz emission from growth optimized InGaAs/InAlAs heterostructures with separated photoconductive and trapping regions. Applied Physics Letters. 2013;**108**(6): 1.4817797

[22] Sartorius B, Roehle H, Künzel H, Böttcher J, Schlak M, Stanze D, et al. All-fiber terahertz time-domain spectrometer operating at 1.5 μm telecom wavelengths. Optics Express. 2008;**16**(13):9565-9570

[23] Ping G, Tani M, Kono S, Sakai K. Study of terahertz radiation from InAs and InSb. Journal of Applied Physics. 2002;**91**:5533

[24] Ascázubi R, Shneider C, Wilke I, Pino R, Dutta PS. Enhanced terahertz emission from impurity compensated GaSb. Physical Review B. 2005;**72**(4):045328

[25] Sigmunda J, Sydlo C, Hartnagel HL. Structure investigation of low-temperature-grown GaAsSb, a material for photoconductive terahertz antennas. Applied Physics Letters. 2005;**87**(25): 1.2149977

[26] Hatem O, Cunningham J, Linfield EH, Wood CD, Davies AG, Cannard PJ, et al. Terahertz-frequency photoconductive detectors fabricated from metal-organic chemical vapor deposition-grown Fe-doped InGaAs. Applied Physics Letters. 2011;**98**(12): 1.3571289

[27] Darrow JT, Zhang X-C, Auston DH, Morse JD. Saturation properties of large-aperture photoconducting antennas. IEEE Journal of Quantum Electronics. 1992;**28**(6):1607-1616

[28] Alfihed S, Foulds IG, Holzman JF. Characteristics of bow-tie antenna structures for semi-insulating GaAs and InP photoconductive terahertz emitters. Sensors. 2021;**21**(9):3131

[29] Pavlović M, Desnica UV. Precise determination of deep trap signatures and their relative and absolute concentrations in semi-insulating GaAs. Journal of Applied Physics. 1997;**84**(4): 1.368258

[30] Alfihed S, Jenne MF, Ciocoiu A, Foulds IG, Holzman JF. Photoconductive terahertz generation in semi-insulating GaAs and InP under the extremes of bias field and pump fluence. Optics Letters. 2021;**46**(3):572-575

[31] Berry CW, Jarrah M. Terahertz generation using plasmonic photoconductive gratings. New Journal of Physics. 2012;**14**:105029

[32] Collier CM, Stirling TJ, Hristovski IR, Krupa JDA, Holzman JF. Photoconductive terahertz generation from textured semiconductor materials. Scientific Reports. 2016;**6**:23185

[33] Lepeshov S, Gorodetsky A, Krasnok A, Toropov N, Vartanyan TA, Belov P, et al. Boosting terahertz photoconductive antenna performance with optimised plasmonic nanostructures. Scientific Reports. 2018;**8**:6624

[34] Georgiou G, Geffroy C, Bäuerle C, Roux J-F. Efficient three-dimensional photonic–plasmonic photoconductive switches for picosecond THz pulses. ACS Photonics. 2020;**7**(6):1444-1451

[35] Tian L, Shi W. Analysis of operation mechanism of semi-insulating GaAs photoconductive semiconductor switches. Journal of Applied Physics. 2007;**103**(12):1.2940728

[36] Kima DS, Citrin DS. Coulomb and radiation screening in photoconductive terahertz sources. Applied Physics Letters. 2006;**88**(16):1.2196480

[37] Binder R, Scott D, Paul AE, Lindberg M, Henneberger K, Koch SW. Carrier-carrier scattering and optical dephasing in highly excited semiconductors. Physical Review B. 1992;**45**:1107-1115

[38] Gorodetsky A, Leite IT, Rafailov EU. Operation of quantum dot based terahertz photoconductive antennas under extreme pumping conditions. Applied Physics Letters. 2021;**119**(11): 5.0062720

[39] Leyman RR, Gorodetsky A, Bazieva N, Molis G, Krotkus A, Clarke E, et al. Quantum dot materials for terahertz generation applications. Laser & Photonics Reviews. 2016;**10**(5):772-779

[40] Mingardi A, Zhang W-D, Brown ER, Feldman AD, Harvey TE, Mirin RP. High power generation of THz from 1550-nm photoconductive emitters. Optics Express. 2018;**26**(11):14472-14478

[41] Singh A, Welsch M, Winnerl S, Helm M, Schneider H. Non-plasmonic improvement in photoconductive THz emitters using nano- and micro-structured electrodes. Optics Express. 2020;**28**(24):35490-35497

[42] Yachmenev AE, Lavrukhin DV, Glinskiy IA, Zenchenko NV, Goncharov YG, Spektor IE, et al. Metallic and dielectric metasurfaces in photoconductive terahertz devices: A review. Optical Engineering. 2019;**59**(6):061608

[43] Yang Y, Kravchenko II, Briggs DP, Valentine J. All-dielectric metasurface analogue of electromagnetically induced transparency. Nature Communications. 2014;**5**:5753

[44] Headley C, Lan F, Parkinson P, Xinlong X, Lloyd-Hughes J, Jagadish C, et al. Improved performance of gaas-based terahertz emitters via surface passivation and silicon nitride encapsulation. IEEE Journal of Selected Topics in Quantum Electronics. 2017;**17**(1):17-21

[45] Gupta A, Rana G, Bhattacharya A, Singh A, Jain R, Bapat RD, et al. Enhanced optical-to-THz conversion efficiency of photoconductive antenna using dielectric nano-layer encapsulation. APL Photonics. 2018;**3**:051706

[46] Chia JY, Tantiwanichapan K, Jintamethasawat R, Sathukarn A. A computational study on performance improvement of THz signal from a grating photoconductive antenna. Photonics. 2020;**7**(4):108

Chapter 2

Studies of Terahertz Sources and Their Applications

Sukhmander Singh, Shravan Kumar Meena, Ashish Tyagi, Sanjeev Kumar, Man Raj Meena and Sujit Kumar Saini

Abstract

The contributed chapter discuss the applications of terahertz radiations and its generation mechanism through laser plasma interactions. The methods of generation of terahertz radiations from plasma wake field acceleration, higher harmonic generation and the laser beat wave plasma frequency are reviewed. The nonlinear current density oscillate the plasma at beat wave frequency under the effect of ponderomotive force and excite the terahertz radiation at beat wave frequency. The current state of the arts of the methods of generation has been incorporated. The mathematical expression of ponderomotive force has been derived under the influence of gradient of laser fields. In additions, the future challenge and their overcomes are also been discussed.

Keywords: electromagnetic waves, THz radiation, beat wave, ponderomotive, nonlinear, plasma, harmonic, detectors

1. Introduction

THz radiation has applications in broadband THz communications, basic science, security, pharmaceutical industries, manufacturing and medicine science. The Terahertz (THz) frequency region, which was difficult accessible frequency region range (0.1–30 THz) lies between the microwave and infrared bands in electromagnetic spectrum. This THz region is also defined as borderline of high frequency region of the microwave band and long wavelength region of far infrared light. Radiation at 1 THz has a period of 1 ps, wavelength 300, wave number 33/cm and photon energy 4.1 meV. It has peak field at 100 MV/cm between 15 THz and 50 THz that provide major momentum to the investigating materials. Therefore THz waves allow direct access to molecular rotations, lattice vibrations and spin waves (low-energy excitations) in contrast to excitations of valence electrons stimulated by optical waves. The THz radiation is non-ionizing and nondangerous for living cells. These radiations can penetrate through plastics, metals, textiles, paper and woods which assists to identify the explosives and drugs. In most of the cases, the vibrational modes of oxygen, water and carbon monoxide (molecules of drugs and explosives) lies in the THz region, therefore during the investigation, those ingredients display distinctive absorption lines in the THz frequency range.

Electromagnetic waves have an applications in medical imaging, broadcasting, WiFi, and treating cancer. The Sequential arrangement of electromagnetic waves are shown in **Table 1**.

We also summarize some of the main applications for each range.

The electromagnetic spectrum

Frequency (Hz)	Nature	Wavelength (m)	Production	Applications
10^{22}	gamma rays	10^{-13}	Nuclear decay	Cosmic rays
10^{21}	gamma rays	10^{-12}	Nuclear decay	Cancer therapy
10^{18}	x rays	10^{-9}	Inner electronic transitions and fast collisions	Medical diagnosis
10^{16}	ultraviolet	10^{-7}		Sterilization
10^{15}	visible	10^{-6}	Thermal agitation and electronic transitions	Vision, astronomy, optical
6.5×10^{14}	blue	4.6×10^{-7}		
5.6×10^{14}	green	5.4×10^{-7}		
3.9×10^{14}	red	7.6×10^{-7}		
10^{14}	infrared	10^{-5}	Thermal agitation and electronic transitions	Heating, night vision, optical communications
10^{9}	UHF	10^{-3}	Accelerating charges and thermal agitation	Microwave ovens
10^{10}	EHF	10^{-1}		remote sensing
10^{8}	TV FM	10		radio transmission
10^{6}	AM	10^{3}		radio signals
10^{4}	RF	10^{5}	Accelerating charges	

Table 1.
The electromagnetic spectrum.

2. Application of terahertz radiation

THz radiation technology have a substantial presentation in the field of engineering, science, biomedical engineering, astrophysics, environmental engineering, information science, technology and plasma physics.

2.1 In biomedicine

Terahertz waves are useful for the diagnosis of disease since every organisms have a unique response to THz wave. THz tomography get absorption rate distribution and three-dimensional distribution of the refractive index of materials in computer assisted tomography [1–29].

2.2 Quality control and safe monitoring

THz radiation is used to observer the process of food processing, weapons, drugs and explosives. THz electromagnetic waves are completely harmless to humans owing to its strong capacity of penetration [2–29].

2.3 Non damaging testing

The penetration length of radiations are measured by THz time-domain spectroscopy. The safety and penetrable properties of THz waves are useful for nondestructive testing. THz waves can penetrate a few inches thick foam. Foam used in space shuttle has very low refractive index and absorption variation, although this change can be observed to detect the defects [1–10].

2.4 Astronomy and atmospheric research

The atmospheric molecules (nitrogen, water, oxygen and carbon monoxide) have excitation energy in the terahertz range, therefore these molecules can be detected with THz radiation to monitor the atmospheric environmental and ozone layer as well as space research [3–12]. THz technology can be employed in astronomy and Earth observation to monitor the weather. The ultra-high frequency of THz radiation provide better digital signal processing and imaging.

2.5 Wireless communication and networking

THz band has higher frequency, wider bandwidth and greater channel than the microwave and 10 Gbps wi-fi transmission speeds may be obtained by means of THz communique, which is some hundred or maybe hundreds of instances faster than contemporary ultra-Wideband technology [3–12].

2.6 Secure communication

The small power of THz radiations is used to gain long-range space communications because of its low attenuation characteristic. THz radiation has wider beam width against space optical communication, which make it suitable to pointing in the long-distance space communication [1–6].

2.7 Chemical and biological agent detection

Terahertz radiation is very sensitive to molecules and surrounding environment. Therefore, terahertz technology is used in chemical detection and identifications of the chemical and biological agents [1–16].

2.8 Medical applications

Ionizing radiation is a kind of radiation that carries photon energy large enough to detach electrons from atoms or molecules, leading to their ionization. X-ray radiation may be harmful for humans. Typical X-ray imaging systems for medical purpose involves photon energies close to 100 keV. Hence, a person may get exposures of a high dose. Since the detection resolution is restricted by diffraction, therefore due to shorter wavelength of the terahertz radiation, it provide better spatial resolution in imaging. Moreover, the vibrational and rotational transition energies of the biomolecular constituents of tissue lies in the THz frequency range, which offer good spectroscopic information of biological tissues. The low photon energy of the radiation is nonionizing and there is negligible scattering from tissues. An exposure of a high dose of ionizing radiation may damage DNA of a human body and may increases the possibility of developing cancer. When the energy is larger than ~10 eV, we say the radiation is ionizing. Therefore the water content presents in the tissue can provides good contrast between the healthy and diseased states of tissues using time-domain spectroscopy based on terahertz radiation. The time-domain spectroscopy provide quasi 3D information in the broad frequency range to investigate the desired information. Although terahertz technology is still young and there have been no major commercial applications in the medical science.

2.9 Quality control and pharmaceutical applications

X-ray photoelectron spectroscopy, Fourier transform infrared and laser induced breakdown spectroscopy are destructive for the medicine tablet to investigate the uniformity of the coating. The non-uniform of the coating or surface defects on the tablets leads to lacks of the desired dose delivery. THz waves have penetrating behavior because of its electromagnetic nature. Terahertz image can be optimized for performing 3D analysis on tablets to determine coating integrity and thickness.

3. Food applications of terahertz spectroscopy

3.1 Terahertz sources

An electronic and photonic materials based methods have been built to generate THz radiation in the recent years and these methods are tabulated in **Table 2**. These sources complement laser-based and other table-top THz sources, which are limited to lower average powers, lower peak fields and lower repetition rates.

Name	Source type	References
Gas	Lasers	Dodel [25]
Semiconductor	Lasers	Chassagneux et al. [26]
Frequency multiplication	Solid-state electronic	Maestrini et al. [27]
Transistors	Solid-state electronic	Lusakowski et al. [28]
Gyrotrons	Vacuum electronic	Bratman et al. [29]
Free electron lasers	Vacuum electronic	Knyazev et al. [30]
Synchrotrons	Vacuum electronic	Byrd et al. [31]
Mercury lamp	Thermal	Charrada et al. [32]
Mechanical resonance	Continuous pumped lasers	Wu et al. [33]
Terahertz parametric oscillator	Pulsed lasers	Kawase et al. [34]

Table 2.
Sources of THz.

4. Ponderomotive force

The nonlinear process arise, when a very high intensity electromagnetic wave interacts with a plasma and the force due to radiation pressure is coupled to the plasma particles and it is called ponderomotive force. Self-focusing of laser light in a plasma is a direct effect of ponderomotive force. When a gas is ionized by propagating a laser thought it, a force exert on the medium at ionization front. A laser beam causes a radially directed ponderomotive force in a plasma which forces plasma out of the beam and dielectric constant becomes higher inside the beam than outside. The plasma acts as a convex lens focusing the beam to a smaller diameter. Here we derive the expression for the ponderomotive force [35].

Equation of motion of particle under the electromagnetic fields

$$m_e \frac{d\vec{v}}{dt} = -e\left[\vec{E}\left(\vec{r}\right) + \vec{v} \times \vec{B}\left(\vec{r}\right)\right] \tag{1}$$

Non-linearity in the system comes partly from the $\vec{v} \times \vec{B}$ term which is a second order term and assuming that $\vec{v_0} = \vec{B_0} = 0$. The other part of nonlinearity comes from evaluating $\vec{E}$ at the actual position of the partical. Let the wave electric field is

$$\vec{E}\left(\vec{r}\right) = \vec{E_S}\left(\vec{r}\right)\cos\left(\omega t\right) \tag{2}$$

We expand E(r) about point $r = r_0$

$$m_e \frac{d\vec{v_1}}{dt} = -e\vec{E}\left(\vec{r_0}\right) = -e\,\vec{E_S}\left(\vec{r_0}\right)\cos\left(\omega t\right) \tag{3}$$

After integrating over time, we get

$$\vec{v_1} = \left(\frac{e}{m_e\omega}\right)\vec{E_S}\left(\vec{r_0}\right)\sin\left(\omega t\right) \tag{4}$$

Again, integrating over time, we get

$$\vec{r_1} = -\left(\frac{e}{m_e\omega^2}\right)\vec{E_S}\left(\vec{r_0}\right)\cos\left(\omega t\right) \tag{5}$$

Now, according to Faraday's law

$$\vec{\nabla} \times \vec{E} = -\frac{\partial \vec{B}}{\partial t} \tag{6}$$

It implies

$$\vec{B_1} = -(1/\omega)\,\vec{\nabla} \times \vec{E} \tag{7}$$

The Taylor expansion of Eq.(2) about point $\vec{r} = \vec{r_0}$

$$\vec{E_S}\left(\vec{r}\right) = \vec{E_S}\left(\vec{r_0}\right) + \left(\vec{r_1} \cdot \vec{\nabla}\right)\vec{E_s}\left(\vec{r} = \vec{r_0}\right) + \ldots\ldots \tag{8}$$

Putting the value in equation in (1) from Eq. (7) and (8), we get

$$m_e \frac{d\vec{v_2}}{dt} = -e\left[\left(\vec{r_1} \cdot \vec{\nabla}\right)\vec{E_s}\left(\vec{r_0}\right)\cos\left(\omega t\right) + \vec{v_1} \times \vec{B_1}\right] \tag{9}$$

On solving the equation and taking the average over time.

$$\left(m_e \frac{d\vec{v_2}}{dt}\right)_{avg} = -(e^2/4m_e\omega^2)\vec{\nabla}\left(E_s{}^2\right)_{avg} \tag{10}$$

Or

$$\left(m_e \frac{d\vec{v_2}}{dt}\right)_{avg} = -(e^2/2m_e\omega^2)\vec{\nabla}\left(E^2\right)_{avg} \tag{11}$$

So the left hand side in equation (11) is the effective force on a single electron, which can be denoted by f_{NLe}.

Similarly force on ions can be written

$$\overrightarrow{f_{NLi}} = \left(m_i \frac{d\overrightarrow{v_2}}{dt}\right)_{avg} = -\left(e^2/2m_i\omega^2\right)\overrightarrow{\nabla}\left(E^2\right)_{avg} \tag{12}$$

So the resultant force on plasma due to ions and electrons is

$$\overrightarrow{F_{rslt}} = n_0\left(\overrightarrow{f_{NLi}} + \overrightarrow{f_{NLe}}\right)$$
$$= \left(\frac{n_{e_0}e^2}{2m_e\omega^2}\right)\left(1 + \frac{m_e}{m_i}\right)\overrightarrow{\nabla}\left(E^2\right)_{avg.} \tag{13}$$

Since the mass of the ions are much greater than the mass of the electrons, we get

$$\overrightarrow{F_{rslt}} = -\left(\omega_{pe}/\omega^2\right)\overrightarrow{\nabla}\left(\varepsilon_0 E^2/2\right)_{avg} \tag{14}$$

This is called Ponderomotive force.

5. Non-linear current due to laser beating

Let us consider two different color laser beams co-propagating in a corrugated plasma having electric field profiles as follows

$$\overrightarrow{E_1} = \hat{j}E_0 e^{-\left(y^2/a_0^2\right)} e^{i(k_1 x - \omega_1 t)} \tag{15}$$

$$\overrightarrow{E_2} = \hat{j}E_0 e^{-\left(y^2/a_0^2\right)} e^{i(k_2 x - \omega_2 t)} \tag{16}$$

The equation of motion of plasma electrons in a field

$$m\frac{\partial \overrightarrow{v}_e}{\partial t} = -e\overrightarrow{E} \tag{17}$$

Lasers impart oscillatory velocity to electrons

$$\overrightarrow{v_1} = e\overrightarrow{E_1}/i\omega_1 m \tag{18}$$

$$\overrightarrow{v_2} = e\overrightarrow{E_2}/i\omega_2 m \tag{19}$$

The corresponding ponderomotive force

$$\overrightarrow{f_p}^{nl} = \frac{-e^2}{2m\omega_i^2}\overrightarrow{\nabla}\left(E_{avg}^2\right) \tag{20}$$

In terms of two components of electric field

$$\overrightarrow{f_p}^{nl} = \frac{-e^2}{2m\omega_1\omega_2}\overrightarrow{\nabla}\left(\overrightarrow{E_1}.\overrightarrow{E_2}^*\right) \tag{21}$$

We know that

$$\overrightarrow{F} = -e\nabla V \tag{22}$$

By comparing the Equation (21) and (22)

$$V = \frac{e}{2m\omega_1\omega_2} \vec{\nabla}\left(\overrightarrow{E_1}.\overrightarrow{E_2}^*\right) \tag{23}$$

the equation (19) deduce that

$$E_i = \frac{im\omega v}{e}. \tag{24}$$

So, from the equation (18) and (19) , we get

$$V = \frac{-m}{2e}\left(\overrightarrow{v_1}.\overrightarrow{v_2}^*\right) \tag{25}$$

From the equation (22)

$$\overrightarrow{f_p}^{nl} = \frac{-e^2}{2m\omega_1\omega_2}\vec{\nabla}\left(\overrightarrow{E_1}.\overrightarrow{E_2}^*\right) \tag{26}$$

Putting the values of electric filed $\overrightarrow{E_1}$ and $\overrightarrow{E_2}$ -

$$\overrightarrow{f_p}^{nl} = \frac{-e^2}{2m\omega_1\omega_2}\vec{\nabla}\left(E_0^2 e^{-\left(2y^2/a_0^2\right)}e^{i[(k_1-k_2)x-(\omega_1-\omega_2)t]}\right) \tag{27}$$

Or

$$\overrightarrow{f_p}^{nl} = \frac{-e^2 E_0^2}{2m\omega_1\omega_2}\vec{\nabla}\left(e^{-\left(2y^2/a_0^2\right)}e^{i\left(k'x-\omega't\right)}\right) \tag{28}$$

Here,

$$k' = (k_1 - k_2) \text{ and } \omega' = (\omega_1 - \omega_2) \tag{29}$$

This oscillatory current is the source for the emission of THz radiation at the beating frequency.

6. Generation of terahertz radiation

A wide range of THz sources are now commercially accessible, although they are big and relatively expensive to run. As a result, much research is being done in order to develop appropriate THz sources. Various organizations across the world have devised various techniques for producing THz sources [5, 6]. Traditional THz sources are based on electro-optic crystals such as ZnSe, GaP, LiNbO3, or photo conductive antennas as well as super-luminous laser pulse interactions with large band gap semiconductors and dielectrics [7–13]. Laser plasma interaction, optical recitation, solid state electronic devices and many complex methods are being used to generate THz radiations [1–20]. The output of such sources can be harmonically multiplied to the THz range. Recent improvement in the field of quantum cascade lasers, laser emission is achieved through the use of inter-sub-band transitions in a periodic repetition of layers of two different compositions, or super-lattice structure. A super-lattice is a periodic structure of quantum wells and barriers. The photon emitted by the super-lattice is due to the intersub-band transition in the super-lattice. Such transitions can be specified by the thickness of the coupled wells and barriers. Therefore, by toiloring the periodicity of the super-lattice to specific well-barrier thickness, THz radiations of specified energy range can be generated. Although the idea of inter-band emission was known since 1971, the crystal growth technology for creating quantum cascade lasers is relatively new and expensive. Terahertz radiation in the frequency range 0.1–10 THz, lies between the MW and IR region and has potential uses in a wide range of fields. That why researchers are interested in this portion of the spectrum [1–5].

7. Schemes based on laser plasma interaction

Terahertz can be generated from nonlinear plasma medium. The following below mentioned schemes are commonly used for the generation of THz radiations.

1. Self-focusing of Laser Beam

2. Wake field Terahertz scheme

3. Beat wave Schemes

The high power laser beams changes the index of refraction of the plasma medium due to non-linear processes, called Self-focusing of laser beam. The increasing intensity of electric field enhances the index of refraction of plasma and the plasma shows similar behaves as a converging lens (**Figure 1**). Further, intensity of self-focusing region rises as the beam enters into medium, until the divergence effect occurs.

A work has been done to investigate terahertz generation in magnetized plasma using self-focusing of hollow Gaussian laser beam [36]. The hollow Gaussian filamented laser propagates parallel to magnetic field and interact with electron plasma wave to produce terahertz radiations. The study shows that intensity of emitted radiations is highly sensitive to the order of hollow Gaussian laser beam. Terahertz generation by amplitude-modulated self-focused Gaussian laser beam in ripple density plasma has also been studied in Ref. [37]. In this system, a current is generated by transverse component of ponderomotive force on electrons as a result the radiation is being driven at the modulation frequency (taken into terahertz domain). It is found that in comparison to without self-focusing to self-focusing an enhancement has been seen in terahertz generation which supported by numerical

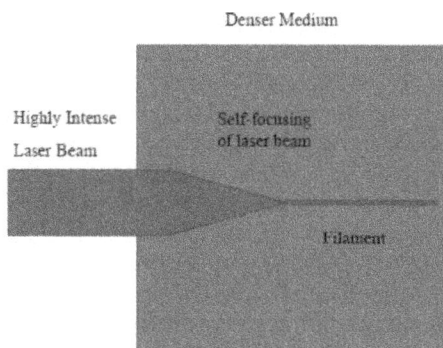

Figure 1.
Self focusing of LASER beam in plasma.

simulation. In rippled density plasma by using cosh-Gaussian lasers terahertz generation has been studied in Ref. [38]. The laser exerts a ponderomotive force along the transverse direction as a result electrons oscillate which get coupled with the density ripple to generate terahertz radiation. It is found that by changing the decentered parameter of laser there exist a notable change in magnitude, amplitude and conversion efficiency of terahertz radiations. Terahertz radiations can also be generated by using relativistic self-focusing hollow Gaussian laser in magnetoplasma [39]. Due to relativistic effect the change in electron mass occurs at high intensity which leads to produce nonlinear effects in plasma leading to the self-focusing of hollow Gaussian laser beam. Hassan et al. defined that when two Laguerre-Gaussian laser beam is gone through the cross-focusing then it generates THz [40]. The amplitude of THz can be enhanced with the help of large amplitude density ripple.

Kumar et al. [41] numerically investigated that THz yield increases sufficiently under the effect of self-focusing and defocusing of amplitude-modulated Gaussian laser beam in rippled density plasma. Hong et al. [42] studied the propagation of a Gaussian and hollow Gaussian laser beam in a tapered plasma to figure out the ponderomotive self-channeling and relativistic self-focusing effects. It has been concluded that, when transverse plasma density is homogeneous, its focusing ability is robust than that of the hollow Gaussian laser. Vhanmore et al. [43] used asymmetric elegant Hermite-cosh-Gaussian to study the self-focusing in magnetized plasma. Kumar et al. [44] analytically studied relativistic self-focusing and particle-in-cell simulations, which reveals that the self-focusing is less sensitive to laser amplitude variation in deeper plasma channels for millimeter range plasma channels present scheme is being valid.

7.1 Wake field terahertz scheme

Laser wake field scheme depends on the interaction of a laser beam with a plasma. A conical emission in the forward direction is produced by laser pulse under the influence of oscillating electrons which involve Cerenkov mechanism is called the laser wake field. Tajima and Dawson explained this scheme [45] as follows. The ponderomotive force originated by laser pulse envelope pushes away the background plasma electrons. The resulting force from the charge separation initiates a density oscillation, after the laser pulse left away the plasma. In this situation, group velocity of the laser equal to the phase velocity of the density oscillation. The same charge fluctuation is known as plasma wave or plasma wake. Self-trapped background plasma electrons produce electron bunch in the wake. Sheng et al. [46] detected powerful coherent emission of terahertz radiation in inhomogeneous plasma, when laser pulse is incident obliquely in laser wake field. It has been observed that the duration of terahertz, frequency and bandwidth depend on laser pulse duration and plasma density profile.

Gupta et al. [47] reported that plasma-density modulation and magnetic field can assist in electron energy enhancement by improving the electron trapping in laser wakefield acceleration (LWFA). Gupta et al. [48] also investigated the acceleration of electrons by the plasma waves in a density rippled inhomogeneous plasma. Gopal and Gupta [49] explored the use of asymmetric laser pulses (of sharp rising front) for optimization and control of electron beam in LWFA and reported that an asymmetric laser pulse reduces the beam emittance, enhance injection and can help in controlling the beam spreading to generate a high-quality monoenergetic beam. Yoshii et al. [50] employed the particle in cell simulation to generate the Cherenkov wake field by a short laser pulse to realize THz radiation in a magnetized plasma. Gopal et al. [51] have also suggested a method of enhancing the magnetic field strength in laser pulse interaction with plasma. Esarey et al. [52] have reviewed the physics of the plasma beat wave accelerator, laser wakefield accelerator and self-modulated laser wakefield accelerator. These sources are capable to handle the strong electric field of order 100 GV/m from an intense laser. Hofmann simulated the performance of quadrupoles and solenoids in focusing and energy selection of laser accelerated protons [53]. Döpp et al. proposed the use of longitudinal density tailoring to reduce the beam chirp at the end of the accelerator [54].

7.2 Beat wave schemes

For the Generation of efficient THz radiation at different frequencies, various experiments have been conducted on lasers beating in a corrugated plasma. THz radiation generated by beating of two lasers yields more tunability and efficiency.
The basic mechanism to generate THz radiation is as follows:
Consider two laser beams having different wave numbers and frequencies propagating in a corrugated plasma. The laser beams exert a ponderomotive force on electrons. As a result these electrons, drives longitudinal oscillations (at beat frequency) adjacent to plasma frequency as shown in **Figure 2**. The generated beat wave decays (parametrically) into a terahertz wave and a plasma wave. The generated terahertz wave is (plasma channel) transverse magnetic mode with finite longitudinal component of the electric field.
Malik et al. [55] used super-posed femtosecond laser pulses to generate the THZ from a gas jet through oscillatory current density. The emission of THz radiation occur through oscillating dipoles. Hamster et al. [56] used 100 femtosecond (1 TW) laser focused onto gas through wakefield. The electrons execute oscillatory motion and produce terahertz radiation under the influences of ponderomotive force. Yampolsky and Frainman [57] reported the four-wave coupling scheme in a plasma filled capillary for the amplification of terahertz radiation. Kukushkin has produced the THz radiations in semiconductors using crossed alternating electric field and static magnetic field [58]. The external dc magnetic field used to increase the field of emitted radiations [59]. Jafari et al. have investi gated that the generation of THz radiation by nonlinear coupling of two color laser beam which have Gaussian field in a plasma with multi-ion species through ponderomotive force on laser in plasma [60]. The radiated THz emission strongly depends on the density of ionic species. Result shows that the maximum value of the amplitude of THz found in a specific range of laser intensity. Li et al.

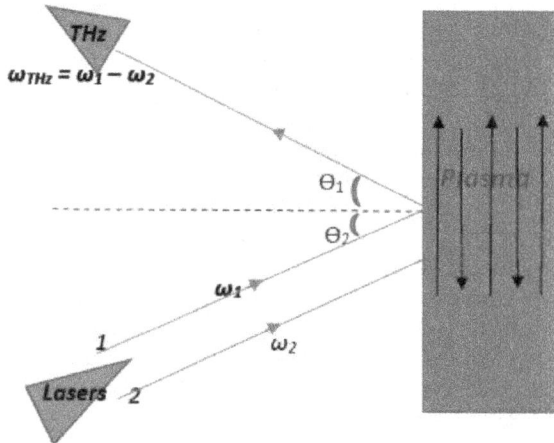

Figure 2.
THZ wave generation in beat wave mechanism.

investigated the contribution of the optical rectification for the generation of THz radiation by two color laser pulse by including the pump power of laser, rotation angle of Beta Barium Borate crystal and the numerical aperture of lens. All of the above factor dramatically affects the intensity of the radiation of THz wave [61]. Bhasin and Tripathi [62] used optical rectification of a x-mode picosecond laser pulse in rippled density magnetized plasma to generate THz radiation. Malik et al. investigated the scheme of the generation of the THz waves with two color laser in clustered plasma. The cluster plasma produces third order nonlinearity and resulted nonlinear current density produce the terahertz radiation. This scheme do not require magnetic field and the density gradient to generate the beat frequencies. The THz conversion efficiency depends on the cluster parameter. It is concluded that the surface plasmon resonance enhances the THz generation efficiency and THz power falls down with THz frequency [63].

In Ref. [64], spatial-Gaussian lasers has been used in a periodic density plasma to excites the radiation which shows depends on the laser-beam width and amplitude. Malik et al. employed two spatial-triangular laser beams for the excitation of terahertz radiation and reported the THz field $\sim 10^5$ kV/cm and the efficiency $\sim 10^{-2}$ correspond to the laser intensity $\sim 10^{14}$ W/cm^2 [65]. Malik and Malik [66] also suggested the mechanism for the generation of tunable terahertz radiation under the application of two femtosecond laser pulses. Dai and Liu [67] studied Terahertz emission in a gaseous plasma (generated by two lasers) with intensity of 5.00×10^{14} W/cm^2. Kumar et al. [68] studied Beat excitation of terahertz radiation from two different frequency infrared lasers of TM/TE mode propagating along z direction in a rippled density semiconductor waveguide slab in a magnetic field (applied transvers to it) and the terahertz yield is significantly higher in the TM mode laser beating than in the TE mode laser beating. Malik and Malik [69–72] investigated the role of an external DC magnetic field in tuning the frequency and power of terahertz radiation. Varshney et al. [73, 74] proposed a scheme for the generation of THz radiation from rippled density magnetized plasma by beating of extraordinary mode lasers.

Malik and Singh [74] used two super-Gaussian lasers to generate the highly focused terahertz radiation by frequency mixing. Chaudhary et al. [75] used Hermite cosh Gaussian lasers to generate the efficient intensity distribution of tunable terahertz radiation. Manendra et al. [76] used hollow sinh super-Gaussian laser beams to generate polarized terahertz wave by photo mixing of two-color laser. The efficiency and the field amplitude increases with electron temperature. Zhang et al. [77] did two-dimensional particle-in-cell simulations of ultra-intense relativistic laser plasma interaction of solid target to generate the terahertz pulses by coherent transition radiation and THz radiation energy increased by 10 times. Manendra et al. [78] investigated the effect of electron temperature on intensity and efficiency of terahertz generated by laser beating in inhomogeneous plasma [79]. Manendra et al. [78–80] used radially polarized lasers having a top-hat envelope profile [s (profile index) ≥ 1] in density modulated hot plasma and concluded that the conversion efficiency increased by 5 times, at the electron thermal velocity 0.2c, where c is speed of light. Liu et al. [81] did two-dimensional particle-in-cell simulations to study the terahertz wave propagating in the stagnation region of a reentry plasma sheath and these investigations are useful to study the attenuation of radio waves in atmosphere communication.

7.3 Resonant third harmonic generation

Higher Harmonic Generation are used to generate a highly coherent radiation sources in the soft x-ray region of the spectrum. When the electric field of the order

10^{13}–10^{14} W/cm^2 of laser interacts with a molecular gas, higher harmonics are produced through nonlinear process laser field. High Harmonic Generation sources has applications in plasma diagnostics, molecular dynamics and in solid state science [82, 83].

Some nonlinear optical crystal describes the formation of the field at the sum frequency of the source fields $\omega 3 = \omega 1 + \omega 2$, where, $\omega 1$, and $\omega 2$ are the frequencies of the sources fields. The crystal produces polarization at a combination of their frequencies and the resultant field oscillate at a frequency $\omega 3$. In second order harmonic generation the resulting field oscillate at 2ω frequency as shown in **Figure 3**.

Cook and Hochstrasser defined that, when we focus the fundamental and second harmonic laser simultaneously whose peak intensities is the order of 10^{14} then it generates the THz radiation [84]. Panwar et al. [85] studied the effect of non-uniform rippled plasma channel on resonant third harmonic laser radiation generation, strongly enhancement of the self-focusing plasma channel non-uniformity and compression of main laser pulse at lower powers and the self-focusing reduces the effectiveness of the third harmonic power because of the compression of main laser in a deeper plasma channel.

Kumar et al. [86] generated the 20 times frequency of the incident wave (high-frequency O-mode radio wave) by nonlinear reflection through ionospheric grating. Kumar and Tripathi [87] studied the parametric coupling of a high amplitude lower hybrid wave with the ion cyclotron instability in tokamak, driven by neutral beam converted ion beam and coupling would be strong when the ion cyclotron wave and the wave numbers of the pump are perpendicular to each other, advanced stage operations of a tokamak as ITER is relevant to it. Kumar [88] investigated the generation of Terahertz radiation by second-order nonlinear mixing of laser and its frequency shifted second harmonic in a rippled density plasma, and emission of THz radiation is maximized when the polarization of the lasers being aligned and also results are accordant with the recent experimental results. Surface Plasmon resonance are used in photonic devices and surface enhanced Raman scattering. Kumar et al. [89] used metal–vacuum of circular surface grating to excite the surface plasma wave and its intensity depends on dimensions of the grating. Tyagi et al. [90] investigated the procedure of third harmonic generation by laser magnetized plasma interaction, and the phase matching condition for the up shifted frequency is satisfied and the laser frequency is not too far from the upper hybrid frequency. Kumar et al. [91] investigated the process of generation of Smith–Purcell terahertz radiation of 10 mW at 10 THz by mixing of two co-propagating lasers passing over a periodic metallic grating.

Kumar and Kumar [92] proposed a scheme of a planar array of nanotube for generation of THz radiation by passing an ultrashort electron bunch. The emitted THz radiation generates at natural frequency of $\omega p/\sqrt{2}$, which is the frequency of electron cylinder.

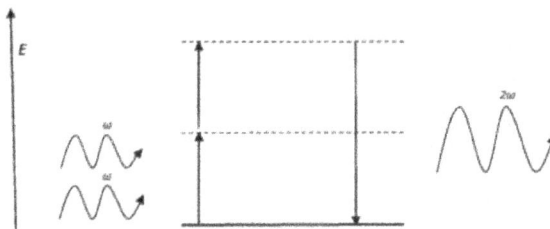

Figure 3.
Mechanics of second order harmonic generation.

It is revealed that the intensity of terahertz radiations is highly sensitive to the magnetic field and the index of super Gaussian beams [93, 94]. By using beating of two super Gaussian lasers in plasma with DC electric field in transverse direction terahertz radiation can be generated [95]. Terahertz radiation generated due to ponderomotive force which acts on electron plasma wave and make them oscillate at frequency difference of two lasers which generates a nonlinear current having frequency in terahertz domain. It is found that the amplitude of terahertz radiation can be enhanced by index of two lasers as well by DC electric field. Investigations are going on for the generation of intense picosecond THz pulses via nonlinear optical methods such as optical rectification. A work has been done to generate terahertz radiations using optical rectification of a super-Gaussian laser beam in rippled density plasma [96]. The change in intensity cause a ponderomotive force in transverse direction which makes electrons oscillates and as a consequence terahertz radiation is being produced. The phase matching is provided by ripple of plasma. In a collision less magnetoplasma terahertz radiation can be produced by using two cross focused Gaussian laser beams [97]. When the applied magnetic field is increased the focusing of lasers increases due to this a nonlinear ponderomotive force acts upon electron plasma waves causing electrons oscillations and a nonlinear current is produced at the terahertz frequency domain. It is found that the amplitude of generated terahertz radiation increases with magnetic field and the cross focusing of two laser beams. The optimization of laser-plasma parameters gives the normalized terahertz power of order 10k. The relativistic focusing of two co-axial Gaussian laser beams into ripple density plasma has been investigated by Kumar et al. [98]. When two lasers propagate into ripple density plasma, then the ponderomotive force reinforce the electrons to oscillates into the transverse direction and these oscillations gets coupled with ripple density of plasma and produce a nonlinear current at terahertz frequency. The study suggest that the amplitude of THz radiation can be enhanced by relativistic ponderomotive focusing of two lasers and also the conversion frequency of the order of 10^{-3} can be achieved. The terahertz generation in collisional plasma using two cross focused laser beams has been studied by Sharma and Singh [99]. The optimized parameters of lasers provide the radiated power of the order 0.23 MW. The applied static electric field, ripple density of plasma and the collision frequency of electron allow the generation of the terahertz radiation. Singh et al. presented a scheme for the generation of strong THz radiation through optical rectification of shaped laser pulse in magnetized plasma [100]. The THz yield increases with the increasing strength of the background magnetic field and the sensitivity depends on the ripple wave number. The emitted power is directly proportional to the square of the amplitude of the density ripple. The enhancement in terahertz generation can be achieved by increasing strength of background magnetic field. It is found that the power of emission is directly proportional to the square of amplitude of ripple. Singh et al. further employed hyperbolic-secant and Gaussian shapes of laser beam to generate the terahertz radiation through optical rectification of a laser pulse in magnetized ripple density plasma. The amplitude of the terahertz radiation shows dependence on the laser beam, laser profile index and the density ripple. When cyclotron frequency approaches to the THz frequency, the THz field amplitude reaches its maximum value [101]. The normalized amplitude of the radiation of order 10^{-2} has been realized. A work has been done to study the generation of terahertz radiations using optical rectification of an amplitude modulated super-Gaussian laser beam propagating into a periodic density plasma with a transverse magnetic field applied on it [102]. The transverse ponderomotive force arises due to the non-uniform spatial variation in laser intensity. The terahertz field amplitude increases with magnetic field strength, modulation index and ripple parameters. Efficiency of the order of 10^{-5} of terahertz wave is achieved. The relativistic ponderomotive force and nonlinear phenomena excite the modulation instability. Jha et al. [103] studied the modulation instability due to the propagation of a laser pulse through a magnetized plasma. It has been depicted that the transverse magnetization of the plasma reduces the modulation instability. In magnetized plasma, the peak spatial growth rate of instability decreased by almost 14 percentage in contrast to the unmagnetized plasma case. Kumar and Tripathi [104] examined Rayleigh scattering of a Gaussian laser beam from clustered gases. According to the model, the clusters expand under laser-induced heating and hydro-dynamic pressure and approach towards plasma resonance. When the cluster electrons reach the plasma frequency of $\sqrt{3}$ times the laser frequency, it produces resonantly enhanced Rayleigh scattering. Magesh and Tripathi [105] investigated the laser excitation of electrostatic eigenmodes of a plasma (having parabolic density profile) in an azimuthal magnetic field. Singh et al. [106] proposed the THz radiation generation by the interaction of the pump upper hybrid wave and the laser (extraordinary wave). In this mechanism, the non-linear interaction between the two waves creates a non-linear current at their frequency difference, which can be brought in the THz range under the appropriate pump frequency and phase matching conditions. In the same research area of THz generation, Hassan et al. [107] studied the interaction of a high-intensity laser beam with density ripple in collisionless magnetized plasma under the paraxial ray approximation to produce THz radiation.

Kumar and Tripathi investigated the schemes of terahertz radiation generation using different methods such as non-linear mixing of laser pulses of finite spot size in clustered gas [108–110], laser bunched electron beam in a magnetic wiggler and optical mixing of laser pulses of finite spot size in a rippled density unmagnetized plasma. In an another study, Rajouria et al. [111] proposed that the relativistic mass and non-linearity increases the resonance absorption of the laser pulse in a density gradient plasma. K K and Tripathi [112] used carbon nanotubes array to investigate the linear and non-linear interaction of laser. The surface plasmon resonance increases, when the laser imparts oscillatory velocity and excursion to electrons in the nanotubes. Kumar et al. [113] have studied the non-linear mixing of laser pulses in a rippled density magnetized plasma. It was obtained that the electron drift induced by lasers' ponderomotive force couples with the density ripple and produces a non-linear current that resonantly drives the THz at the beat frequency. Liu et al. [114], developed an analytical formalism for broadened surface plasmon resonance and enhanced X-ray emission is achieved in a non-uniform clusters with high power lasers. Kumar et al. [115] explored the laser beat wave excitation of THz radiation in a hot plasma with a step density profile, where enhanced yield is achieved due to the coupling with the Langmuir wave at plasma frequency near THz frequency.

Bakhtiari et al. [116] proposed a scheme for improving terahertz radiation efficiency by the interaction of two Gaussian laser array beams in an electron-neutral collisional plasma. They optimized that high efficiency of up to 0.07% can be achieved using array beams, which is almost three times higher than the maximum efficiency achieved by a single Gaussian laser beam. An analytical study has been presented by Sharma et al. [117] for the generation of terahertz radiation due to transverse wakefields produced by the propagation of a circularly polarized laser pulse in a homogeneous, underdense, and axially magnetized plasma. Sobhani et al. [118] demonstrated the vital role of pump depletion and cross-focusing effects in the generation of twisted THz radiation in a non-linear plasma medium. Lehmann and Spatschek [119] discussed the generation of plasma gratings in underdense plasma by counterpropagating laser pulses, which can act as plasma photonic crystals for high-power lasers. In Ref. [10, 120] review of recent progresses in the generation, detection and application of intense terahertz radiation has been reported.

8. Detection of terahertz pulses

The electro-optic sampling is used to detect the THZ radiations which is opposite of optical rectification mechanism in nonlinear crystals. Coherent detection process is normally used so that the amplitude and phase of radiation can be detected. Other way of detection of Terahertz wave is electro optic effect. This detection method is based on the process in which electric field at terahertz frequency induce a birefringence in an optically transparent material. The relation between the magnitudes of effect is directly proportional to the state of the field. Fourier transform of the temporal pulse give the THz spectrum.

The direct and coherent detectors are mainly used to identify the terahertz radiations. The direct detector measure the average power and the coherent detectors measure the instantaneous value of electric field.

8.1 Direct detectors

The Bolometer, Golay cell and the pyroelectric are used to measure the average power of broadband THz pulses. The bolometer work on the principle of temperature-dependent electrical resistivity which contains of a sensing material. As soon as it absorb the incoming photons, its shows the change of its electrical resistivity when it is illuminated by incoming radiations.

8.2 Coherent detectors

The atto-second technologies are used to measure the amplitude and phase of the electric field oscillation of an EM wave at THz frequency. Although this technique is difficult at optical frequencies, therefore it is achieved in the radio wave frequency range with the oscilloscope.

The THz radiation photon have energies of the order of few meV than the photons of optical frequencies. Therefore the ambient background and thermal noise disturb the measurement of THz radiations. So, it has become necessary to extract the background noise from the interested signals. The researcher use the Phase Sensitive Detection linked with an instrument called Lockin Amplifier.

9. Challenges in the terahertz radiation spectroscopy

There are various challenges in the field of terahertz radiation spectroscopy and imaging. The first issue is that terahertz radiation are strongly absorbed by the polar liquid (water) which presents in all the tissues, so they cannot penetrate much deeper into the moist tissues [1, 3, 4].

The other challenges in the terahertz radiation spectroscopy and imaging are resolution and its slowness mechanism in comparison to previously established ways of imaging, which produces thousands of pixels per second while the terahertz have the speed 1 pixel in several seconds. For any conventional way of imaging the diffraction is limited by wavelength of the radiation. In the case of terahertz imaging, it lies in the range of one micrometer to 3 mm which do not give enough detail images for most of the medical applications. This shortcoming can be overcome by near field imaging.

10. Conclusion

This chapter contribute the applications and generations of THz technology in the field of security, medicines, science and biomedical engineering. The biomacromolecules and certain drugs are detected by using of THz spectroscopy. THz imaging has been employed to diagnostics of cancers, treatment skin burn and dental related diagnostics. The biological effect of THz is still required to further explore the research area. The nonthermal effects of THz radiation on human DNA is needed to pay more attention. For the widespread applications of THz, we need higher-power THz sources and their THz detectors.

Author details

Sukhmander Singh[1]*, Shravan Kumar Meena[2], Ashish Tyagi[3], Sanjeev Kumar[2], Man Raj Meena[4] and Sujit Kumar Saini[2]

1 Plasma Waves and Electric Propulsion Laboratory, Department of Physics, Central University of Rajasthan, Ajmer, Rajasthan, India

2 Department of Physics, Motilal Nehru College, University of Delhi, Delhi, India

3 Department of Physics, Swami Shraddhanand College, University of Delhi, Delhi, India

4 Department of Physics, Gargi College, University of Delhi, New Delhi, India

*Address all correspondence to: sukhmandersingh@curaj.ac.in

IntechOpen

References

[1] Lewis RA, Lewis RA. Terahertz physics. Cambridge: Cambridge University Press; 2012

[2] Perenzoni M, Paul DJ. Physics and Applications of Terahertz Radiation. New York: Springer; 2014

[3] Bründermann E, Hübers HW, Kimmitt MF. Terahertz Techniques. Heidelberg: Springer; 2012

[4] Saeedkia D. Handbook of Terahertz Technology for Imaging, Sensing and Communications. Woodhead, Philadelphia, Elsevier; 2013

[5] Lewis RA. A review of terahertz sources. Journal of Physics D: Applied Physics. 2014;**47**(37):374001

[6] Wilmink GJ, Grundt JE. Invited review article: current state of research on biological effects of terahertz radiation. Journal of Infrared, Millimeter, and Terahertz Waves. 2011; **32**(10):1074-1122

[7] Tonouchi M. Cutting-edge terahertz technology. Nature Photonics. 2007; **1**(2):97-105

[8] Siegel PH. Terahertz technology in biology and medicine. IEEE Transactions on Microwave Theory and Techniques. 2004;**52**(10): 2438-2447

[9] Hu BB, Nuss MC. Imaging with terahertz waves. Optics Letters. 1995; **20**(16):1716-1718

[10] Zhang Y, Li K, Zhao H. Intense terahertz radiation: Generation and application. Frontiers of Optoelectronics. 2020;**23**: 1-33

[11] Lewis RA. A review of terahertz detectors. Journal of Physics D: Applied Physics. 2019;**52**(43):433001

[12] Wei F. Review of terahertz semiconductor sources. Journal of Semiconductors. 2012;**33**(3):031001

[13] Crowe TW, Porterfield DW, Hesler JL, Bishop WL, Kurtz DS, Hui K. Terahertz sources and detectors. In Terahertz for Military and Security Applications III. International Society for Optics and Photonics. 2005;**5790**: 271-280

[14] Davies AG, Linfield EH, Johnston MB. The development of terahertz sources and their applications. Physics in Medicine & Biology. 2002;**47**(21): 3679

[15] Zhong K, Shi W, Xu D, Liu P, Wang Y, Mei J, et al. Optically pumped terahertz sources. Science China Technological Sciences. 2017;**60**(12): 1801-1818

[16] Nagatsuma T. Terahertz technologies: present and future. IEICE Electronics Express. 2011;**8**(14): 1127-1142

[17] Mukhopadhyay SJ, Hazra P, Mitra M. A brief review on terahertz avalanche transit time sources. Advanced Materials for Future Terahertz Devices, Circuits and Systems. 2021;**727**:43

[18] Ding YJ. Progress in terahertz sources based on difference-frequency generation. Journal of the Optical Society of America B. 2014;**31**(11): 2696-2711

[19] Fice MJ, Rouvalis E, Ponnampalam L, Renaud CC, Seeds AJ. Telecommunications technology-based terahertz sources. Electronics Letters. 2010;**46**(26):28-31

[20] Fülöp JA, Tzortzakis S, Kampfrath T. Laser-driven strong-field terahertz

sources. Advanced Optical Materials. 2020;**8**(3):1900681

[21] Lee YS. Principles of terahertz science and technology. Berlin: Springer Science & Business Media; 2009

[22] Ferguson B, Zhang XC. Materials for terahertz science and technology. Nature Materials. 2002;**1**(1):26-33

[23] Zhang XC, Xu J. Introduction to THz Wave Photonics. New York: Springer; 2010

[24] Kim KW, Kim H, Park J, Han JK, Son JH. Terahertz tomographic imaging of transdermal drug delivery. IEEE Transactions on Terahertz Science and Technology. 2011;**2**(1):99-106

[25] Dodel G. On the history of far-infrared (FIR) gas lasers: Thirty-five years of research and application. Infrared Physics & Technology. 1999; **40**:127-139

[26] Chassagneux Y, Colombelli R, Maineult W, Barbieri S, Beere HE, Ritchie DA, et al. Electrically pumped photonic-crystal terahertz lasers controlled by boundary conditions. Nature. 2009;**457**:174-178

[27] Maestrini A, Ward JS, Gill JJ, Lee C, Thomas B, Lin RH, et al. A frequency-multiplied source with more than 1mW of power across the 840–900-GHz band. IEEE Transactions on Microwave Theory Technology. 2010;**58**: 1925-1932

[28] Lusakowski J et al. Voltage tuneable terahertz emission from a ballistic nanometer InGaAs/InAlAs transistor. Journal of Applied Physics. 2005;**97**: 064307

[29] Bratman VL, Kalynov YK, Manuilov VN. Large-orbit gyrotron operation in the terahertz frequency range. Physical Review Letters. 2009;**102**:245101

[30] Knyazev BA, Kulipanov GN, Vinokurov NA. Novosibirsk terahertz free electron laser: Instrumentation development and experimental achievements Meas. Science and Technology. 2010;**21**:054017

[31] Byrd JM, Leemans WP, Loftsdottir A, Marcelis B, Martin MC, McKinney WR, et al. Observation of broadband self-amplified spontaneous coherent terahertz synchrotron radiation in a storage ring. Physical Review Letters. 2002;**89**:224801

[32] Charrada K, Zissis G, Aubes M. Two-temperature, two-dimensional fluid modelling of mercury plasma in high-pressure lamps. Journal of Physics D: Applied Physics. 1996;**29**:2432-2438

[33] Wu XL, Xiong SJ, Liu Z, Chen J, Shen JC, Li TH, et al. Green light stimulates terahertz emission from mesocrystal microspheres. Nature Nanotechnology. 2011;**6**:103-106

[34] Kawase K, Sato M, Taniuchi T, Ito H. Coherent tunable THz-wave generation from $LiNbO_3$ with monolithic grating coupler. Applied Physics Letters. 1996;**68**:2483-2485

[35] Chen FF. Introduction to Plasma Physics and Controlled Fusion. 3rd edition. Switzerland: Springer Nature; 2015

[36] Hussain S, Singh M, Singh RK, Sharma RP. THz generation by self-focusing of hollow Gaussian laser beam in magnetised plasma. EPL. 2014;**107**: 65002

[37] Kumar S, Singh RK, Singh M, Sharma RP. THz radiation by amplitude-modulated self-focused Gaussian laser beam in ripple density plasma. Laser and Particle Beams. 2015; **33**:0263-0346

[38] Singh M, Singh RK, Sharma RP. EPL. 2013;**104**:35002

[39] Hussain S, Singh RK, Sharma RP. Laser and Particle Beams. 2016;**34**:86-93

[40] Sobhani H, Dehghan M, Dadar E. Effect of pump depletion and cross-focusing on twisted terahertz radiation generation. Physics of Plasmas. 2017; **24**(2):023110

[41] Kumar S, Singh RK, Singh M, Sharma RP. THz radiation by amplitude-modulated self-focused Gaussian laser beam in ripple density plasma. Laser and Particle Beams. 2015; **33**(2):257-263

[42] Hong X-R, Zheng Y-L, Tang R-A, Liu T-F, Liu X-P. Physics of Plasmas. 2020;**27**:043109

[43] Vhanmore BD, Takale MV, Patil SD. Physics of Plasmas. 2020;**27**:063104

[44] Kumar A, Dahiya D, Sharma AK. Laser prepulse induced plasma channel formation in air and relativistic self focusing of an intense short pulse. Physics of Plasmas. 2011;**18**(2): 023102

[45] Tajima T, Dawson JM. Laser electron accelerator. Physical Review Letters. 1979;**43**(4):267

[46] Sheng ZM, Mima K, Zhang J, Sanuki H. Emission of electromagnetic pulses from laser wakefields through linear mode conversion. Physical Review letters. 2005;**94**(9):095003

[47] Gupta DN et al. Laser wakefield acceleration of electrons from a density-modulated plasma. Laser and Particle Beams. 2014;**32**(3):449-454

[48] Gupta DN et al. Mode-coupling assisted electron accelerations by a plasma wave. Current Applied Physics. 2015;**15**(3):174-179

[49] Gopal K, Gupta DN. Optimization and control of electron beams from laser wakefield accelerations using

asymmetric laser pulses. Physics of Plasmas. 2017;**24**(10):103101

[50] Yoshii J, Lai CH, Katsouleas T, Joshi C, Mori WB. Radiation from Cerenkov wakes in a magnetized plasma. Physical Review Letters. 1997;**79**(21):4194

[51] Gopal K, Gupta DN, Kim YK, Hur MS, Suk H. Large-scale magnetic field generation by asymmetric laser-pulse interactions with a plasma in low-intensity regime. Journal of Applied Physics. 2016;**119**(12):123101

[52] Esarey E, Schroeder CB, Leemans WP. Physics of laser-driven plasma-based electron accelerators. Reviews of Modern Physics. 2009;**81**(3): 1229

[53] Hofmann I. Performance of solenoids versus quadrupoles in focusing and energy selection of laser accelerated protons. Physical Review Special Topics-Accelerators and Beams. 2013;**16**(4):041302

[54] Döpp A, Thaury C, Guillaume E, Massimo F, Lifschitz A, Andriyash I, et al. Energy-chirp compensation in a laser wakefield accelerator. Physical Review Letters. 2018;**121**(7):074802

[55] Malik AK, Malik HK, Kawata S. Journal of Applied Physics. 2010;**107**: 113105

[56] Hamster H, Sullivan A, Gordon S, Falcone RW. Short-pulse terahertz radiation from high-intensity-laser-produced plasmas. Physical Review E. 1994;**49**(1):671

[57] Yampolsky NA, Frainman GM. Physics of Plasmas. 2006;**13**:113108

[58] Kukushkin VA. Generation of THz radiation in semiconductors with cyclotron heating of heavy holes. EPL (Europhysics Letters). 2008;**84**(6): 60002

[59] Hamster H, Sullivan A, Gordon S, White W, Falcone RW. Subpicosecond, electromagnetic pulses from intense laser-plasma interaction. Physical Review Letters. 1993;**71**(17):2725

[60] Jafari MJ, Jafari Milani MR, Rezaei S. Terahertz radiation from multi ion plasma irradiated by two cross focused Gaussian laser beams. Physics of Plasmas. 2019;**26**(10):103107

[61] Li H, Zhang Y, Sun W, Wang X, Feng S, Ye J, et al. Contribution of the optical rectification in terahertz radiation driven by two-color laser induced plasma. Optics Express. 2020;**28**(4):4810-4816

[62] Bhasin L, Tripathi VK. Physics of Plasmas. 2009;**16**:103105

[63] Malik R, Uma R, Kumar P. Two color laser driven THz generation in clustered plasma. Physics of Plasmas. 2017;**24**(7):073109

[64] Malik AK, Malik HK, Nishida Y. Physics Letters A. 2011;**375**:1191-1194

[65] Malik AK, Malik HK, Stroth U. Applied Physics Letters. 2011;**99**:071107

[66] Malik HK, Malik AK. Applied Physics Letters. 2011;**99**:251101

[67] Dai H, Liu J. Photonics and Nanostructures – Fundamentals and Applications. 2012;**10**:191-195

[68] Kumar M, Bhasin L, Tripathi VK. Journal of Physics and Chemistry of Solids. 2012;**73**:269-274

[69] Malik AK, Malik HK, Stroth U. Physical Review E. 2012;**85**:016401

[70] Malik AK, Malik HK. IEEE Journal of Quantum Electronics. 2013;**49**:2

[71] Malik AK, Singh KP, Sajal V. Physics of Plasmas. 2014;**21**:073104

[72] Malik AK, Singh KP. Laser and Particle Beams. 2015;**33**:519-524

[73] Varshney P, Sajal V, Singh KP, Kumar R, Sharma NK. Laser and Particle Beams. 2013;**31**:337-344

[74] Varshney P, Sajal V, Chauhan P, Kumar R, Navneet KS. Laser and Particle Beams. 2014;**32**:375-381

[75] Chaudhary S, Singh KP, Pal B, Manendra S, Malik AK. Modeling of intense terahertz wave generation with controlled field distribution. Physics of Plasmas. 2019;**26**:073107

[76] Manendra SC, Singh KP, Sheoran G, Malik AK. A Letters Journal Exploring the Frontiers of Physics. 2019;**126**:55001

[77] Zhang S, Yu J, Shou Y, Zheng G, Li D, Geng Y, et al. Physics of Plasmas. 2020;**27**:023101

[78] Manendra KPS, Singh BP, Malik AK. Investigation of effect of electron temperature on intensity and efficiency of terahertz generated by laser beating in inhomogeneous plasma. Physica Scripta. 2020;**95**(11):115007

[79] Singh MKP, Bhati R, Malik AK. Efficient terahertz (THz) generation by nonlinear mixing of bicolor top-hat lasers in hot plasma. Physics of Plasmas. 2020;**27**:023108

[80] Singh MKP, Singh BP, Malik AK. Physics of Plasmas. 2020;**27**:063101

[81] Liu J-x, Zhao Y, Lv J-j, Shi Q, Liu T-y, Yu T-p, et al. AIP Advances. 2021;**11**:065001

[82] Cavalieri AL, Müller N, Uphues T, Yakovlev VS, Baltuška A, Horvath B, et al. Attosecond spectroscopy in condensed matter. Nature. 2007;**449**:1029

[83] Theobald W, Häßner R, Wülker C, Sauerbrey R. Temporally resolved

measurement of electron densities
($>1023 cm^{-3}$) with high harmonics.
Physical Review Letters. 1996;77:298

[84] Cook DJ, Hochstrasser RM. Intense
Terahertz Pulses by Four-Wave
Rectification in Air. Optics Letters;
2000;25(16);1210-1212

[85] Panwar A, Ryu CM, Kumar A.
Effect of plasma channel non-
uniformity on resonant third harmonic
generation. Laser and Particle Beams.
2013;31(3):531-537

[86] Ashok Kumar RU, Tripathi VK.
Radio Science. 2006;41:RS4014

[87] Kumar A, Tripathi VK. Parametric
coupling of a lower hybrid pump with
neutral beam driven ion cyclotron
instability in a tokamak. Physics of
Plasmas. 2008;15:062509

[88] Kumar A. Ponderomotive self-
focusing of surface plasma wave.
Plasmonics. 2013;8:1135-1139

[89] Pawan Kumar VKT, Kumar A, Shao
X. Launching focused surface plasmon
in circular metallic grating. Journal of
Applied Physics. 2015;117:013103

[90] Tyagi Y, Tripathi D, Kumar A.
Bernstein wave aided laser third
harmonic generation in a plasma.
Physics of Plasmas. 2016;23:093115

[91] Kumar P, Bhasin L, Tripathi VK,
Kumar A, Kumar M. Smith–Purcell
terahertz radiation from laser
modulated electron beam over a
metallic grating. Physics of Plasmas.
2016;23:093301

[92] Kumar A, Kumar P. Electron beam
induced THz emissions from nanotube
array. Physics of Plasmas. 2016;23:
103302

[93] Jha P et al. Second harmonic
generation in laser magnetized–plasma

interaction. Physics of Plasmas. 2007;
14(5):053107

[94] Kumar M, Tripathi VK. Nonlinear
absorption and harmonic generation of
laser in a gas with anharmonic clusters.
Physics of Plasmas. 2013;20(2):023302

[95] Hussain S, Singh RK, Sharma RP.
Terahertz radiation generation by
beating of two super Gaussian lasers in
plasma having static dc electric field.
Physics of Plasmas. 2016;23(7):073120

[96] Kumar S, Singh RK, Sharma RP.
Strong terahertz generation by optical
rectification of a super-Gaussian laser
beam. EPL. 2016;114:55003

[97] Singh RK, Sharma RP. Terahertz
generation by two cross focused
Gaussian laser beams in magnetized
plasma. Physics of Plasmas. 2014;21:
113109

[98] Kumar S, Singh RK, Sharma RP.
Terahertz generation by relativistic
ponderomotive focusing of two co-axial
Gaussian laser beams propagating in
ripple density plasma. Physics of
Plasmas. 2015;22:103101

[99] Sharma RP, Singh RK. Terahertz
generation by two cross focused laser
beams in collisional plasmas. Physics of
Plasmas. 2014;21:073101

[100] Singh RK, Rajoria MSSK, Sharma
RP. Strong terahertz emission by optical
rectification of shaped laser pulse in
transversely magnetized plasma.
Physics of Plasmas. 2017;24:073114

[101] Singh RK, Rajoria MSSK, Sharma
RP. High power terahertz radiation
generation by optical rectification of a
shaped pulse laser in axially magnetized
plasma. Physics of Plasmas. 2017;24:
103103

[102] Singh M, Kumar S, Singh RK, Uma
R, Sharma RP. High-power terahertz
emission in magnetized plasma via

optical rectification of a super-Gaussian laser beam. EPL. 2017;**119**:15002

[103] Jha P et al. Modulation instability of laser pulse in magnetized plasma. Physics of Plasmas. 2005;**12**(12):123104

[104] Kumar M, Tripathi VK. Rayleigh scattering of a Gaussian laser beam from expanding clusters. Physics of Plasmas. 2009;**16**(12):123111

[105] Kumar KKM, Tripathi VK. Laser excitation of electrostatic eigenmode of a plasma in azimuthal magnetic field and electron acceleration. Physics of Plasmas. 2010;**17**(5):053103

[106] Singh M, Kumar S, Sharma RP. Terahertz wave generation by the upper hybrid wave. Physics of Plasmas. 2011; **18**(2):022304

[107] Hassan MB et al. Terahertz generation by the high intense laser beam. Journal of Plasma Physics. 2012; **78**(5):553-558

[108] Manoj K, Tripathi VK. Terahertz generation by nonlinear mixing of laser pulses in a clustered gas. Physics of Plasmas. 2011;**18**(5):053105

[109] Manoj K, Tripathi VK. Terahertz radiation from a laser bunched relativistic electron beam in a magnetic wiggler. Physics of Plasmas. 2012;**19**(7): 073109

[110] Manoj K, Tripathi VK. Resonant terahertz generation by optical mixing of two laser pulses in rippled density plasma. IEEE Journal of Quantum Electronics. 2012;**48**(8):1031-1035

[111] Rajouria SK, Magesh Kumar KK, Tripathi VK. Nonlinear resonance absorption of laser in an inhomogeneous plasma. Physics of Plasmas. 2013;**20**(8): 083112

[112] Magesh Kumar K, Tripathi VK. High power laser coupling to carbon

nano-tubes and ion Coulomb explosion. Physics of Plasmas. 2013;**20**(9):092103

[113] Kumar M, Rajouria SK, Magesh Kumar KK. Effect of pulse slippage on beat wave THz generation in a rippled density magnetized plasma. Journal of Physics D: Applied Physics. 2013; **46**(43):435501

[114] Liu CS, Tripathi VK, Kumar M. Interaction of high intensity laser with non-uniform clusters and enhanced X-ray emission. Physics of Plasmas. 2014; **21**(10):103101

[115] Kumar M, Tripathi VK, Jeong YU. Laser driven terahertz generation in hot plasma with step density profile. Physics of Plasmas. 2015;**22**(6):063106

[116] Bakhtiari F et al. Terahertz radiation generation and shape control by interaction of array Gaussian laser beams with plasma. Physics of Plasmas. 2016;**23**(12):123105

[117] Sharma P, Wadhwani N, Jha P. Terahertz radiation generation by propagation of circularly polarized laser pulses in axially magnetized plasma. Physics of Plasmas. 2017;**24**(1):013102

[118] Sobhani H, Dehghan M, Dadar E. Effect of pump depletion and cross-focusing on twisted terahertz radiation generation. Physics of Plasmas. 2017; **24**(2):023110

[119] Lehmann G, Spatschek KH. Laser-driven plasma photonic crystals for high-power lasers. Physics of Plasmas. 2017;**24**(5):056701

[120] Hafez HA, Chai X, Ibrahim A, Mondal S, Férachou D, Ropagnol X, et al. Intense terahertz radiation and their applications. Journal of Optics. 2016;**18**(9):093004

Section 2

System Dynamics

Chapter 3

Research and Application of PID Controller with Feedforward Filtering Function

Biao Wang and Shaojun Lin

Abstract

Most of the existing differential methods focus on the differential effect and do not make full use of the differential link's filtering effect of reducing order and smoothing. In Proportion Integral Differential (PID) control, the introduction of differential can improve the dynamic performance of the system. However, the actual differential (containing differential gain) will be subject to the impact of high-frequency noises. Therefore, this paper proposes a differential with filtering function, which has weak effect on noise amplification, and strong effect on reducing order and smoothing. Firstly, a discrete differentiator was constructed based on the Newton interpolation, and the concept of "algorithm bandwidth" was defined to ensure the differential effect. Then, the proposed algorithm was used to design a new PID controller with feedforward filtering function. In the experiments, the proposed PID controller is applied to a high-performance hot water supply system. The result shows that the system obtains better control quality. It verifies that the proposed PID controller has a feedforward filtering function and can effectively remove high-frequency noise.

Keywords: Newton interpolation, algorithm bandwidth, the discrete differentiator, PID controller, feedforward filtering

1. Introduction

As Proportion Integral Differential (PID) control is widely used in industrial control, engineering applications, and other fields [1, 2], a problem in Proportion Integral Differential (PID) control cannot be ignored: the introduction of differential signals can improve the dynamic characteristics of the system, but it is also easy to introduce high-frequency interference [3], the insufficiency of the differential term is especially obvious when the error perturbation is abruptly changed. The feedforward control can generate a compensation amount in advance according to the magnitude of the disturbance after the disturbance occurs and before the controlled variable has changed, thereby eliminating the influence of the disturbance on the controlled variable. Therefore, a filter or feedforward link is generally added to the PID control system to improve the system performance, but this makes the system structure more complicated.

The existing differentiation method focuses on the differential function of the differential link, but involves less on the filtering function of the differentiator.

Although the differential link has the function of order reduction, harmonics will appear at the same time as the order reduction. If it is not smoothed, the control effect is not ideal when applied to the actual control system. Therefore, it is necessary to design a PID controller with filtering function that has weak effect on noise amplification and strong reduction and smoothing effect.

In the current methods of extracting input differential signals, wavelet [4] and neural network [5] rely on system models or may amplify noise. In recent years, sliding mode algorithms have been used to design differentiators or filters, but there is a problem of chattering elimination [6–8]. Literature [9] realized a dither-free sliding mode differentiator, which constructed discrete differential based on backward Euler, but the algorithm parameter adjustment is more complicated. Tracking differentiator (TD) can quickly track the input signal [3], but its more parameters increase the difficulty of application in practice [10, 11]. The structure of fractional differential is generally more complicated. For example, the fractional differential in literature [12] is implemented with a complex combination structure of low-pass, high-pass, and band-pass filter functions.

Numerical differentiation [13–16] approximates the function value of the unknown objective function at certain points based on the information of the known finite discrete sampling points. According to the literature, numerical differentiation mainly includes: finite difference type [17], polynomial interpolation type [13, 15], regularization method [18, 19], and undetermined coefficients [20]. Among them, the commonly used method is finite difference, but its effect is not very ideal for the high-frequency noise existing in the measurement process [21].

In view of the above problems, due to the advantages of Newton interpolation that it is convenient to calculate a large number of interpolation points, this paper constructs a "discrete differentiator" based on the equidistant Newton interpolation polynomial [22]. The proposed differentiator can realize the effect of differential and filter, so it can replace the strategy of "controller and filter" in some application. Compared with the existing Gaussian filtering [23], Kalman filtering [23–25], and Wiener filtering [25] schemes, the proposed discrete differentiator has a simple structure, and it is easy to implement.

Then this article gives a simulation example, using the discrete differentiator as the differential link in PID control, applied to the hot water temperature supply system, and compared with the control effect of the feedforward and feedback compound control (using the traditional PID controller) strategy. The results show that the PID controller simplifies the structure of the system. Even if it does not have the feedforward link, it still has the feedforward filtering function. It overcomes the negative factors of the differential link in the traditional PID controller that amplify the noise and produce side effects. The system has obtained better control quality.

2. A numerical differential algorithm based on Newton's interpolation

The distribution of sampling points for many problems in practical engineering is often equidistant. Eq. (1) shows that the algorithm based on the equally spaced forward-difference Newton's interpolation polynomials only utilizes the previous sampling points. Therefore, the "on-line" control can be achieved. Namely, the differential of the point y_0 is estimated from the target-function's values of $y_0, y_{-1}, y_{-2}, \ldots, y_{-k}$ at the moment of the current period t_0 and previous periods $t_0 - \Delta t, t_0 - 2\Delta t, \ldots, t_0 - k\Delta t$.

$$\hat{y}(u) = y_0 + u\Delta y_{-1} + \frac{1}{2!u(u+1)\Delta y_{-2}} + \frac{1}{3!u(u+1)(u+2)\Delta y_{-3}} + \cdots \quad (1)$$

Eq. (1) gives the Newton's interpolation function based on equal interval, and we introduce variable $u = (t - t_0)/\Delta t$ (Δt is the sample step) to simplify the interpolation function.

When $t = t_0 (u = 0)$, the first derivative of the Newton's interpolation polynomial has the form in Eq. (2):

$$\hat{y}'(t_0) = \frac{1}{\Delta t}\left(\Delta y_{-1} + \frac{\Delta^2 y_{-2}}{2} + \frac{\Delta^3 y_{-3}}{3} + \frac{\Delta^4 y_{-4}}{4} + + \frac{\Delta^5 y_{-5}}{5} + \frac{\Delta^6 y_{-6}}{6} + \frac{\Delta^7 y_{-7}}{7} + \ldots \right)$$

(2)

Δ^m is forward differential operator, Eq. (2) provides various forms of Δ^m:

$$\Delta y_{-1} = y_0 - y_{-1}$$

$$\Delta^2 y_{-2}/2 = \frac{1}{2}\left(y_0 - 2y_{-1} + y_{-2}\right)$$

$$\Delta^3 y_{-3}/3 = \frac{1}{3}\left(y_0 - 3y_{-1} + 3y_{-2} - y_{-3}\right)$$

$$\Delta^4 y_{-4}/4 = \frac{1}{4}\left(y_0 - 4y_{-1} + 6y_{-2} - 4y_{-3} + y_{-4}\right)$$

(3)

$$\Delta^5 y_{-5}/5 = \frac{1}{5}\left(y_0 - 5y_{-1} + 10y_{-2} - 10y_{-3} + 5y_{-4} - y_{-5}\right)$$

$$\Delta^6 y_{-6}/6 = \frac{1}{6}\left(y_0 - 6y_{-1} + 15y_{-2} - 20y_{-3} + 15y_{-4} - 6y_{-5} + y_{-6}\right)$$

The numerical differential algorithm formulas involving two to seven sampling points can be obtained from Eq. (2). **Table 1** lists the equally spaced forward-difference formula based on the first derivative of the Newton's interpolation.

The closed-loop system's bandwidth is determined by the cutoff frequency ω_c on the amplitude-frequency characteristics of the open-loop system. This band range is necessary to ensure that the numerical differential algorithm's accuracy meets the system requirements. Therefore, this paper defined the concept of the "algorithm bandwidth." It refers a range in which the frequency characteristics of the ideal differential and the proposed differentiator are basically consistent or the maximum frequency limit when the differential accuracy of the ideal differential and the proposed method are basically equal.

The algorithm bandwidth is calculated by comparing the frequency characteristics of the proposed method and the ideal differential. Therefore, we can get two

Points number	The forward-difference formula
2	$\hat{y}'(t_0) = \frac{1}{\Delta t}(\Delta y_{-1}) = \frac{y_0 - y_{-1}}{\Delta t}$
3	$\hat{y}'(t_0) = \frac{3y_0 - 4y_{-1} + y_{-2}}{2\Delta t}$
4	$\hat{y}'(t_0) = \frac{11y_0 - 18y_{-1} + 9y_{-2} - 2y_{-3}}{6\Delta t}$
5	$\hat{y}'(t_0) = \frac{25y_0 - 48y_{-1} + 36y_{-2} - 16y_{-3} + 3y_{-4}}{12\Delta t}$
6	$\hat{y}'(t_0) = \frac{137y_0 - 300y_{-1} + 300y_{-2} - 200y_{-3} + 75y_{-4} - 12y_{-5}}{60\Delta t}$
7	$\hat{y}'(t_0) = \frac{147y_0 - 360y_{-1} + 450y_{-2} - 400y_{-3} + 225y_{-4} - 72y_{-5} + 10y_{-6}}{60\Delta t}$

Table 1.
The equally spaced forward-difference formula with 2–7 sampling points.

values of algorithm bandwidth, which are respectively in the sense of amplitude-frequency characteristics and phase-frequency characteristics.

For the convenience of description, the following contents of this paper will replace the numerical differential algorithm with the differentiator filter. For a differentiator filter when two sampling points are adopted, we will replace it with the two-point differentiator filter.

The standard for judging the effect of the proposed differentiator mainly considers the following three aspects:

- Accuracy of differentiation in algorithm bandwidth;

- Maximum transmission coefficient outside the range;

- Easy implementation;

The most significant is the transmission coefficient of differentiator filter. When a differentiator filter is similar to an ideal differential in the high-frequency characteristic, it would amplify high-frequency noises. Nevertheless, this factor is undesirable when it occurs in practical applications.

In section III, we compared the proposed differentiator with the ideal differential and the actual differential by analyzing their frequency characteristics. Besides, the frequency characteristics of the above differentiators are calculated by Laplace Fourier transforms.

3. The discrete differentiator

3.1 The design of discrete differentiator

For the ideal differential, its transfer function and amplitude-frequency characteristics, and phase-frequency characteristics are in Eq. (4).

$$W(j\omega) = K_d \bullet j\omega$$
$$|W(j\omega)| = K_d \bullet \omega$$
$$\varphi(\omega) = +\pi/2 \tag{4}$$

K_d in Eq. (4) is the transfer coefficient of the differential link. Without losing generality, we suppose that $K_d = 1$.

A differentiator filter can be constructed by Eq. (2) using a first derivative of the Newton's interpolation polynomials. Eq. (5) gives the two-point differentiator:

$$\hat{y}'(n\Delta t) = \frac{[y(n\Delta t) - y((n-1)\Delta t)]}{\Delta t} = f(n\Delta t) \tag{5}$$

Applying Laplace transform and Fourier transforms to Eq. (5), we can get transfer function of the two-point differentiator in Eq. (6).

$$W^*(j\omega\Delta t) = \frac{F^*(j\omega)}{Y^*(j\omega)} = \frac{(1 - \exp(-j\omega\Delta t))}{\Delta t}$$

$$= \frac{[1 - \cos(\omega\Delta t) + j \bullet \sin(\omega\Delta t)]}{\Delta t} \tag{6}$$

Substituting the real part and the virtual part of Eq. (6) into Eq. (7) and the frequency characteristics of the two-point differentiator filter will be obtained:

$$L(\omega) = 20 \bullet \lg \left(\frac{\sqrt{\text{Re}^2 + \text{Im}^2}}{\Delta t} \right)$$

$$\varphi(\omega) = \text{arctg} \left(\frac{\text{Im}}{\text{Re}} \right) \tag{7}$$

The amplitude-frequency characteristics and the phase-frequency characteristics of the two-point differentiator filter are given in Eq. (8) and Eq. (9) by calculating Eq. (7). The calculation process is not difficult, so I will not explain it in detail here.

$$|W^*(j\omega\Delta t)| = 2 \bullet \frac{\sin\left(\frac{\omega\Delta t}{2}\right)}{\Delta t} \approx \omega \tag{8}$$

$$\varphi(\omega\Delta t) = \text{arctg} \left[\frac{\sin(\omega\Delta t)}{(1 - \cos(\omega\Delta t))} \right]$$

$$= \text{arctg} \left[\frac{\text{ctg}(\omega\Delta t)}{2} \right] \tag{9}$$

It should be noted that the amplitude-frequency characteristics of the proposed method and the ideal differential should be basically the same.

The algorithm bandwidth of the two-point differentiator filter is the frequency range when the frequency characteristics of the two-point differentiator and the ideal differential are consistent within a specific precision range. From now on, algorithm bandwidth also refers to this upper-frequency limit.

In terms of the amplitude-frequency characteristics, we can get the estimation of algorithm bandwidth in Eq. (10) through Eq. (8). That is the algorithm bandwidth of the two-point differentiator filter in the sense of amplitude-frequency characteristics.

$$\omega \leq \frac{1}{\Delta t} \tag{10}$$

The phase of ideal differential is $+\pi/2$, so the phase of a two-point differentiator filter should be $+\pi/2$, too. Making Eq. (9) $\varphi(\omega\Delta t) = \text{arctg}[\text{ctg}(\omega\Delta t)/2] = +\pi/2$, we can get the algorithm bandwidth in the sense of phase-frequency characteristics in Eq. (11). Note that the ideal solution of $\varphi(\omega\Delta t) = \text{arctg}[\text{ctg}(\omega\Delta t)/2] = +\pi/2$ is 0 and Eq. (11) is an approximate solution to satisfy the above relationship.

$$\omega \leq \frac{0.1}{\Delta t} \tag{11}$$

Similarly, the frequency characteristics and algorithm bandwidth of the n-point differentiator ($n =2,3,4,5,6,7$) based on the first derivative of the Newton's interpolation can be obtained by using the method described above. The transfer function of the three-point discrete differentiator to the seven-point discrete differentiator is shown in **Table 2**.

3.2 The frequency characteristics of discrete differentiator

Next, we will compare the proposed differentiator with ideal differential and the actual differential by analyzing their frequency characteristics. We give the

Points number	$W^*(j\omega)$
2	$\dfrac{1-e^{-j\omega\Delta t}}{\Delta t}$
3	$\dfrac{3-4e^{-j\omega\Delta t}+e^{-2j\omega\Delta t}}{2\Delta t}$
4	$\dfrac{11-18e^{-j\omega\Delta t}+9e^{-2j\omega\Delta t}-2e^{-3j\omega\Delta t}}{6\Delta t}$
5	$\dfrac{25-48e^{-j\omega\Delta t}+36e^{-2j\omega\Delta t}-16e^{-3j\omega\Delta t}+3e^{-4j\omega\Delta t}}{12\Delta t}$
6	$\dfrac{137-300e^{-j\omega\Delta t}+300e^{-2j\omega\Delta t}-200e^{-3j\omega\Delta t}+75e^{-4j\omega\Delta t}-12e^{-5j\omega\Delta t}}{60\Delta t}$
7	$\dfrac{147-360e^{-j\omega\Delta t}+450e^{-2j\omega\Delta t}-400e^{-3j\omega\Delta t}+225e^{-4j\omega\Delta t}-72e^{-5j\omega\Delta t}+10e^{-6j\omega\Delta t}}{60\Delta t}$

Table 2.
The transfer function of the three-point discrete differentiator to the seven-point discrete differentiator.

simulation of the proposed method with various sampling points and the sample step Δt to analyze the effect of the sampling points and the sample step on a differentiator filter.

We give the simulation of the amplitude-frequency characteristics and phase-frequency characteristics of the differentiator filters and ideal differential in **Figure 1**. The differentiator filters respectively adopt 2, 3, 4 sample points and their sample step are all 0.1 second. Therefore, transfer functions of the two-discrete differentiator, the three-discrete differentiator and the four-discrete differentiator are $\frac{1-e^{-jw\cdot 0.1}}{0.1}$, $\frac{3-4e^{-jw\cdot 0.1}+e^{-jw\cdot 0.2}}{2*0.1}$, $\frac{11-18e^{-jw\cdot 0.1}+9e^{-jw\cdot 0.2}-2e^{-jw\cdot 0.3}}{6*0.1}$.

From the result, it can be seen that the cutoff frequency ω_c of the ideal differential and the discrete differentiators is 1 rad/s. The ideal differential is a linear function of +20 dB per ten times the frequency, and the output amplitude increases with the increase of input frequency. Therefore, it would amplify the noise in the high-frequency region because its amplitude increases with the increase of the frequency. Whereas amplitude of the proposed differentiators with 2, 3, 4 sample points does not increase with the increase of the frequency, and it is finally stable. The four-point discrete differentiator has a 36.2 dB maximum amplitude, three-point discrete differentiator has a 32 dB maximum amplitude, two-point discrete differentiator has a 26.9 dB maximum amplitude.

We obtain $\omega \leq 1/\Delta t = 10$rad/s and $\omega \leq 0.1/\Delta t = 1$rad/s ($\Delta t = 0.1$s) by Eq. (10). Namely, for the two-point differentiator, algorithm bandwidth in the sense of amplitude-frequency characteristics and phase-frequency characteristics is respectively 10 rad/s and 1 rad/s. It can be seen in **Figure 1**, the amplitude-frequency characteristic of the proposed method is consistent with the ideal differential in the region of 0 rad-10 rad/s, and the phase-frequency characteristic of the proposed method is consistent with the ideal differential in the region of 0 rad–1 rad/s which confirms Eq. (10).

3.3 Influence of parameters the sampling step Δt and sampling points n on the filtering effect of the discrete differentiator

In order to study the Influence of parameter the sampling step Δt on the filtering effect of the discrete differentiator, we give some simulations. **Figure 2** shows the frequency characteristics of the two-point differentiator filter based on the sample step $\Delta t = 0.1$s and $\Delta t = 1$s.

In **Figure 2**, the cutoff frequency ω_c of the discrete differentiators is 1 rad/s. In the high-frequency region, the blue curve has a smaller amplitude compared with the green curve. The two-point discrete differentiator with $\Delta t = 0.1$s has a 26.9 dB

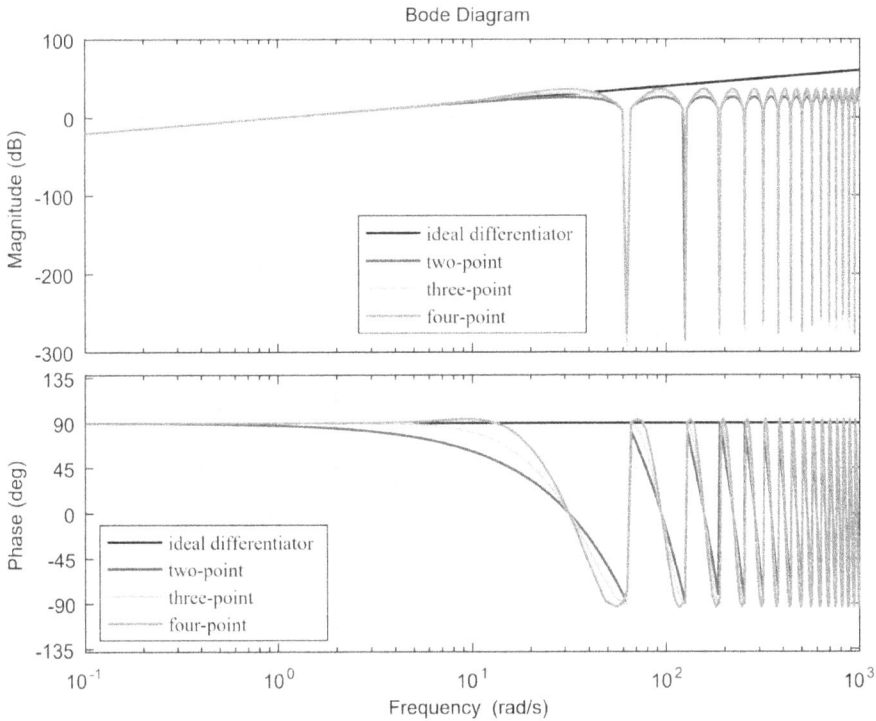

Figure 1.
Frequency characteristic of ideal differential and the differentiator filters when sample step is 0.1 s ($\Delta t = 1s$).

maximum amplitude, and the two-point discrete differentiator with $\Delta t = 1s$ has a 5.5 dB maximum amplitude. It indicates that when the sample step Δt takes the larger value, the differentiator filter becomes weaker for the high-frequency noise amplification. Therefore, we should take Δt (the sampling time) as large as we can, but Δt could not be too large, or it would affect the accuracy and stability of the differentiator.

Figures 3 and **4** shows the Nyquist diagram of the differentiator filter with 2–7 points differentiator filter. We can infer that the high-frequency transmission coefficient increases as the number of sampling points in the algorithm increases. Furthermore, the two-point differentiator has the maximum error and minimum transmission coefficient in the high-frequency region. Oppositely, the seven-point differentiator has the minimum error in algorithm bandwidth and the maximum transmission coefficient in the high-frequency region.

Besides, in **Figures 3** and **4**, when the number of sampling points exceeds 4, it is observed that the root locus of the differentiator filter appears on the left side of the **s** domain. The differentiator filter constructed at this time does not conform to the characteristics of the differential link. In addition, adopting too many sampling points will increase the algorithm complexity.

Therefore, the proposed differentiator takes up to four sampling points. It is not recommended to increase the number of sampling points for differentiator filters.

As the ideal differential could not be synthesized in practical scenes, the ideal differential does not exist. The actual differential generally has certain inertia so it can be realized in series with the ideal differential. Another essential feature of discrete differentiation is that the maximum transfer coefficient (gain) of the

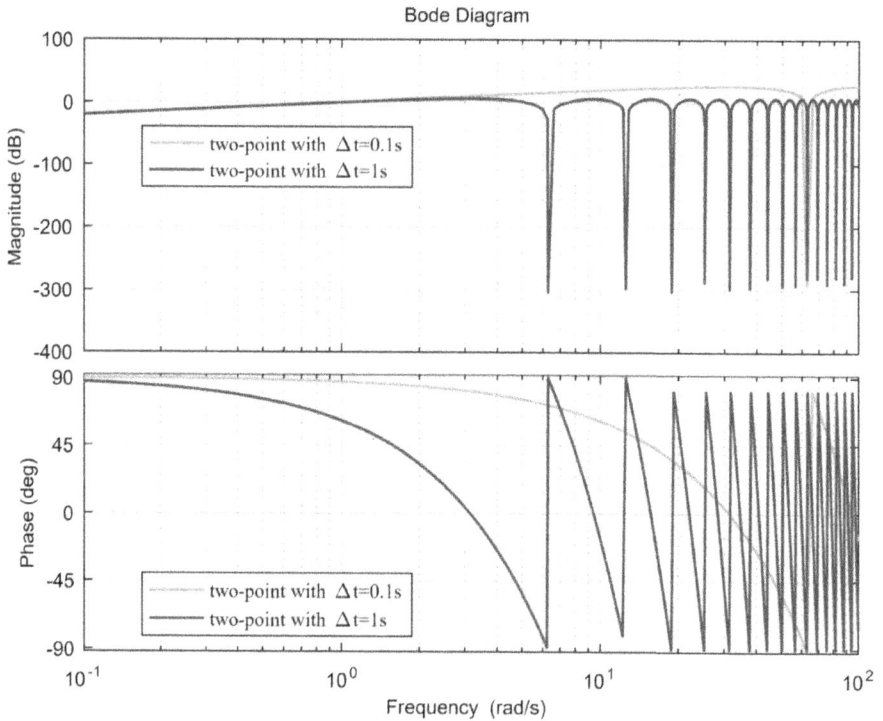

Figure 2.
Frequency characteristic of the two-point differentiator filters when sample step is 1 s ($\Delta t = 1$s) and 0.1 s ($\Delta t = 0.1$s).

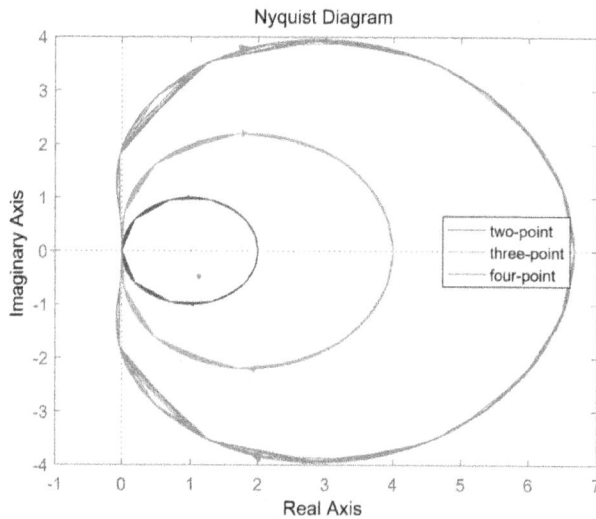

Figure 3.
Root locus of the proposed differentiator with 2–4 points.

differentiator is subject to the bandwidth limit. Eq. (12) gives transfer function of an actual differential.

$$W_d(j\omega) = \frac{j\omega}{(1 + T_d j\omega)} \qquad (12)$$

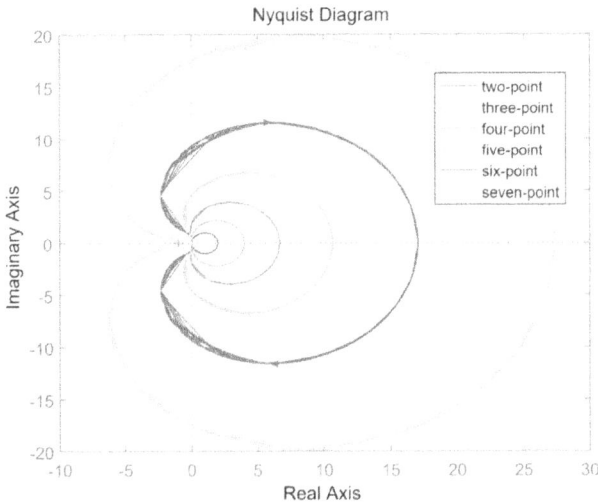

Figure 4.
Root locus of the proposed differentiator with 2–7 points.

T_d- the equivalent time constant of the actual differential link. We can obtain T_d by the following relationship: the modulus of the actual differential link in Eq. (12) should be equal to that of the differentiator filter when $\omega \to \infty$.

Eq. (13) gives the transfer function of two-point differentiator filters, three-point differentiator filters, and four-point differentiator filters:

$$W^*(j\omega) = \frac{(1 - \exp(-j\omega\Delta t))}{\Delta t}$$

$$W^*(j\omega) = \frac{(3 - 4\exp(-j\omega\Delta t) + \exp(-2j\omega\Delta t))}{2\Delta t}$$

$$W^*(j\omega) = \frac{(11 - 18\exp(-j\omega\Delta t) + 9\exp(-2j\omega\Delta t) - 2\exp(-3j\omega\Delta t))}{6\Delta t} \tag{13}$$

While $\omega \to \infty$, the modulus of the actual differential is as follows:

$$|W_d(j\omega)| = \frac{|\omega|}{\sqrt{1 + (T_d\omega)^2}} = \frac{\omega}{(T_d\omega)} = 1/T_d \tag{14}$$

For the two-point differentiator filter, its modulus can be got as the following relation when $\omega \to \infty$:

$$|W^*(j\omega\Delta t)| = \frac{2}{\Delta t} \bullet \sin\left(\frac{\omega\Delta t}{2}\right) = \frac{2}{\Delta t} \tag{15}$$

Make Eq. (14) equal to Eq. (15), we can get Eq. (16):

$$\frac{2}{\Delta t} = \frac{1}{T_d} \tag{16}$$

Calculate the solution of Eq. (16) and get Eq. (17):

$$T_d = \frac{2}{\Delta t} \tag{17}$$

In the same way, we can find the relation between T_d and Δt when the actual differential link corresponds to the three-point differentiator filter and four-point differentiator filter.

$$T_d = \frac{\Delta t}{k}, \qquad (k = 4, 6.667) \tag{18}$$

In order to evaluate the effect of differentiator filters more effectively, the frequency characteristic of the differentiator filter is compared with the frequency characteristic of the actual differential link.

The differentiating time constant of the actual differential link is 0.25 s, which corresponds to the three-point differentiator filter when $\Delta t = 1$s, the frequency characteristics of both are given in **Figure 5**.

It can be seen from the figure that the two-point discrete differentiator and the actual differential have the same maximum of 12 dB amplitude, so the proposed discrete differentiator can realize the effect of suppressing noise like the actual differentiator. The continuous simulation of the discrete algorithm in the high-frequency domain can be analyzed on the transfer coefficient of the differentiator filter so that the possible noise on the signal transmission path after the control device can be determined to a great extent.

With the improvement of microprocessor control equipment performance, the existing control equipment has high sampling frequency and high processing speed. Implementing discrete differential algorithms on these devices can achieve the same effect as continuous algorithms. When the differential link of the PID controller is realized by the differentiator filter and the continuous differentiator, the frequency characteristics of the two are close.

While constructing the differentiator filter Δt through the first derivative method of Newton's interpolation, we find it that the disturbance suppression effect of the differentiator filter on the high-frequency noise is more evident when the number of sampling points is smaller, and the sample step is larger. The differentiator filter has the best high-frequency noise suppression ability when two sampling points are selected, and the maximum allowable sample step Δt is selected. It should be noted that when the selected sampling points are small, the differential accuracy of the constructed differentiator filter will be reduced.

The standard deviation estimates of the input and output random signals of the actual and differentiator filters under different noise are given in **Table 3**. Where the sample step of the differentiator filter is 1 second ($\Delta t = 1$s). The equivalent time constant T_d of the actual differentiator is $1/k (T_d = 1/k)$.

In **Table 3**, comparing with the actual differential, the output of the proposed method had smaller standard deviation under different noise. There are three common types of noise: normal distribution noise, white noise, and uniform distribution noise.

Normal distribution noise also calls Gaussian noise, which is a kind of noise whose probability density function obeys Gaussian distribution. Common Gaussian noise includes fluctuation noise, cosmic noise, thermal noise, shot noise, and so on. The three noises are very common, and it is convenient to quantify and evaluate the effect of the algorithm.

White noise is a random signal with constant power spectral density. In other words, the power spectral density of this signal is the same in each frequency band. Because white light is mixed with monochromatic light of various frequencies (colors), the property of this signal with flat power spectrum is called "white," and this signal is also called white noise.

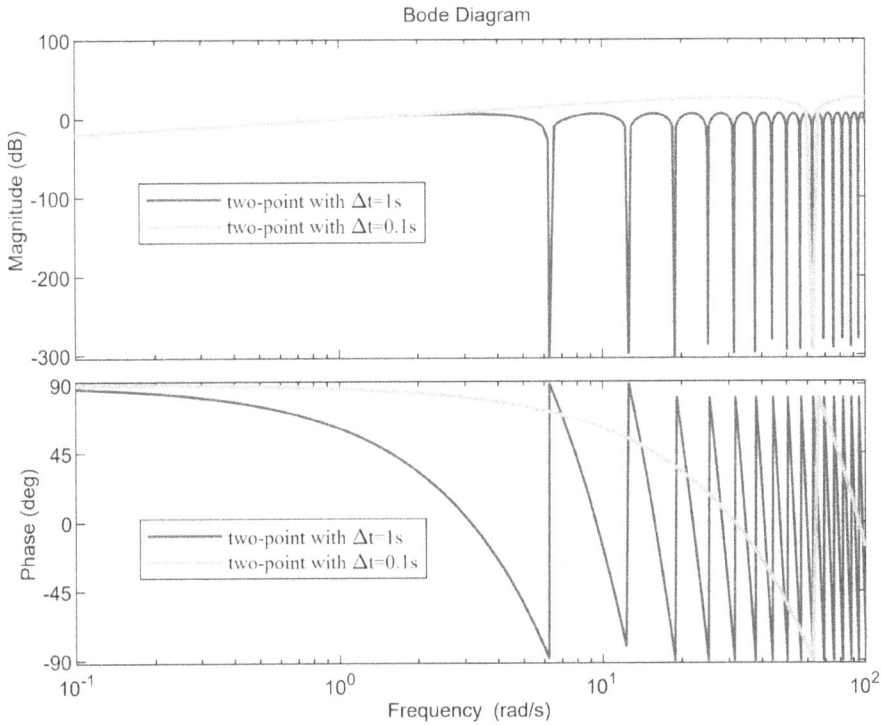

Figure 5.
Frequency characteristics of the actual differential when $T_d = 0.25$s and the three-point differentiator filter when $\Delta t = 1$s.

Noise	Input	Actual differential	The proposed differentiator		
			2	3	4
Normal distribution signal	0.98	1.99	1.249		
		3.79		2.29	
		6.48			3.39
White noise	3.32	6.39	4.18		
		12.15		7.34	
		19.06			10.72
Uniform distribution signal	1.81	3.51	2.286		
		6.52		4.47	
		11.50			5.00

Table 3.
Standard deviation (SD) of the actual differential and the proposed method.

Uniformly distributed noise is a kind of noise whose probability density function obeys uniform distribution. Uniform distribution is one of the important distributions in probability and statistics.

We give the algorithm effect under these three noises in order to verify the adaptability and robustness of the algorithm. It shows that the algorithm can deal with them effectively for common noises and the differential accuracy of the differentiator filter is higher than that of the actual differential, and the differential effect is better.

In combination with the previous article, it is concluded that when taking the maximum allowable sample step Δt, the two-point differentiator, three-point differentiator, and four-point differentiator filter based on the Newton's interpolation have higher differential accuracy in the algorithm bandwidth and smaller transmission coefficient in the high-frequency region when compared with the actual differentiator. Therefore, the differentiator filter can effectively suppress the high-frequency interference.

4. The experiment and result analysis

We apply the proposed method in the high-performance hot water supply systems. The system belongs to the closed-loop system, and the controlled variable is temperature. The system consists of the following parts: comparator, PID controller, three-position relay, actuator, heat exchanger, and temperature sensor. The transfer function and parameters of the high-performance hot water supply system are also given in **Figure 6**.

In **Figure 6**, the three-position relay is a relay with dead zone characteristics. The dead zone width is determined by the deviation of the actual water temperature from the water temperature setting value.

Then we should consider the method of adjusting the parameters of the PID controller. In the proposed algorithm, the differential link in the PID controller is realized by the differentiator filter and the actual differential. Next, the specific method of adjusting PID controller parameters is introduced.

To reduce the system's sensitivity to high-frequency noise, the differentiator filter in the PID controller would adopt an increased sample step Δt, which is calculated by Eq. (10).

For an actual differential whose transfer function is $W_d(s) = k_d s/(T_d + 1)$, according to Eq. (18), T_d is got by Eq. (11). Moreover, k = 2, 4, 6.667, respectively corresponds to the coefficient of two-point differentiator filters, three-point differentiator filters, and four-point differentiator filters.

After adopting the ideal differential in PID controller, a filter is added (T_s of the filter is the same as the time constant T_d of the inertia link), and the effect of the control system is the same as that of the PID controller using the actual differential.

For the convenience of description, PID1 refers to the PID controller designed by the actual differentiator, and PID2 refers to the PID controller designed by the discrete differentiator with 2–4 points. Besides, the only difference between PID1 and PID2 is their differential link that PID1 adopted the actual differential link and PID2 adopted the proposed discrete differential link.

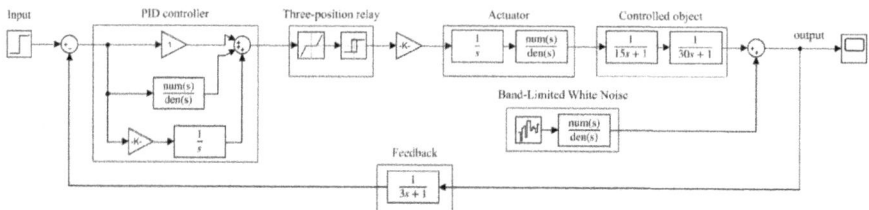

Figure 6.
The high-performance hot water supply system.

Table 4 shows the relative error of the modulus and phase of the system when the PID controller adopted the differentiator filter with 2–4 points and the actual differentiator at the cutoff frequency of the system, which are compared with the ideal differentiator.

Table 4 indicates that the phase-frequency characteristics of the proposed differentiators have the smaller relative error and slight distortion near the cutoff frequency, which illustrates that it can meet the system requirements for transient quality.

Then we give several simulations to analyze the effect of the proposed PID controller in the high-performance hot water supply system, where the system applied PID1 and PID2 respectively. **Figure 7** shows the amplitude-frequency and phase-frequency characteristics of the PID controllers of the high-performance hot water supply system, where the differential link structure of the PID controllers is defined as follows:

1. PID1-PID controller with the actual differential link.

2. PID2-PID controller with the discrete differential with an increased sample step.

The PID controllers in above were applied in high-performance hot water supply system. The results of the system's frequency characteristics were shown in **Figure 8**.

From **Figures 7** and **8**, it can be seen that the system with the differentiator filters has the more remarkable effect of high-frequency attenuation than the system with the actual differentiators near the frequency $\omega_n = n \cdot \frac{2\pi}{\Delta t}, n = 1, 2, 3 \ldots$, where n is an integer.

Next, the PID control strategy with the proposed differentiator and the feedforward-feedback control strategy were adopted in the above high-performance hot water supply system. **Figure 9** shows the simulation structure of the second strategy: feedforward-feedback control system. In which, the selected noise is a 2 KHz pulse signal with amplitude 5, and the PID controller is a traditional PID in the form of continuous time. The pulse signal was selected as system noise instead of the band-limited white noise because feedforward cannot overcome unmeasurable noise. Besides, the pulse signal can be regarded as sudden interference such as industrial interference caused by high-frequency electrical equipment and hot water consumption in the hot water supply system, which can represent the actual random system interference. In order to highlight the effectiveness of the proposed differentiator, the noise of the scheme in **Figure 6** keeps up with feedforward-feedback control strategy.

The relative error	PID1	PID2		
		2	3	4
Modulus	9.0%	9.37%		
	5.0%		1.0%	
	3.0%			0.008%
Phase	7.0%	7.0%		
	4.0%		1.0%	
	2.0%			0.001%

Table 4.
The relative error of the differentiator filter with 2–4 sampling points and the actual differentiator.

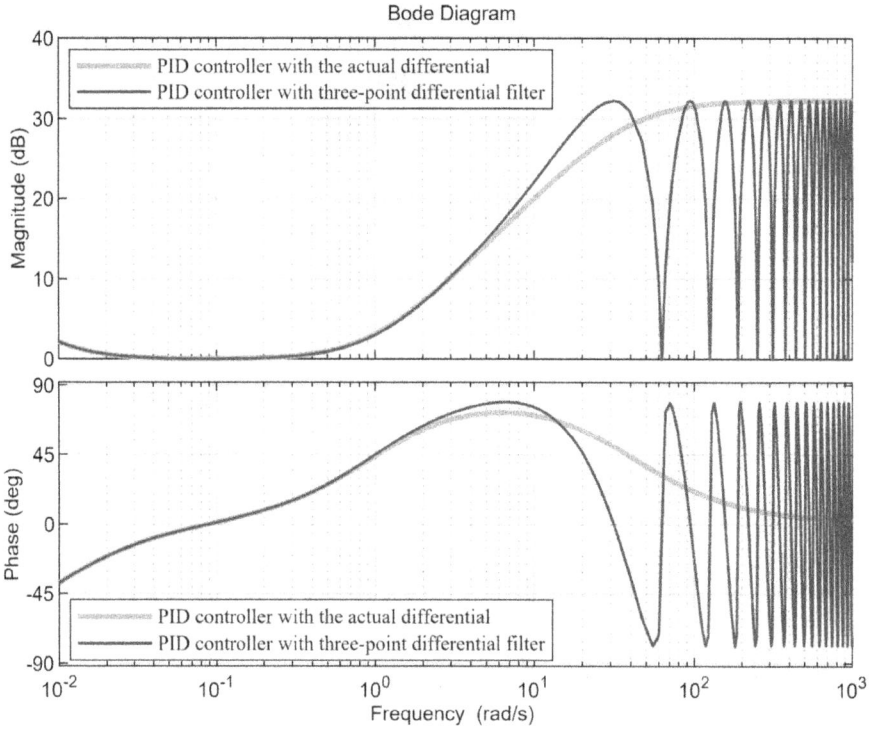

Figure 7.
Frequency characteristics of PID controller with the actual differential and the three-point differentiator.

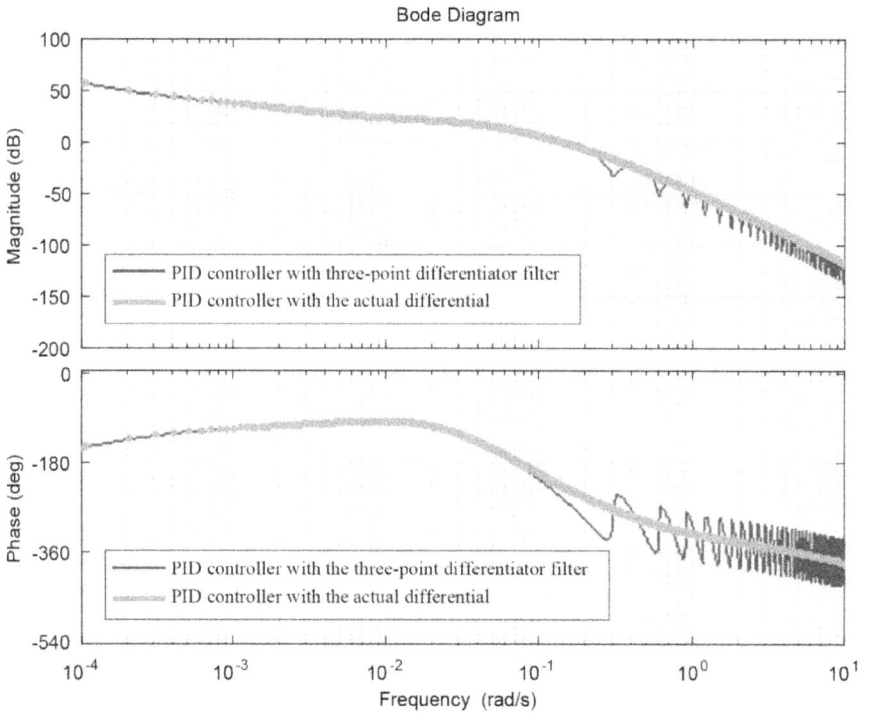

Figure 8.
Frequency characteristics of open-loop hot water supply system.

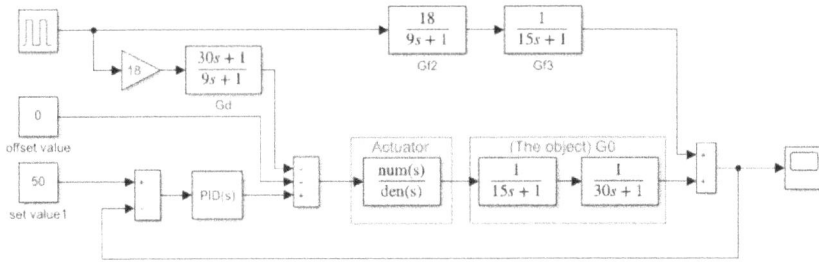

Figure 9.
The simulation structure of the feedforward-feedback control system.

As mentioned above, PID 1 refers to the PID controller with the actual differential and PID2 refers to the PID controller with the discrete differentiator. The comparison simulations of the PID control strategy with the proposed differentiator, the actual differentiator and the feedforward-feedback control strategies are given in **Figure 10**. PID1 adopted the three-point discrete differentiator with $\Delta t = 1s$. The parameters in PID1 are $K_p = 3, K_i = 1, K_d = 0.05$. The parameters in PID2 are $K_p = 3, K_i = 0.5, K_d = 0.05$.

It can be seen from the figure that the algorithm has less overshoot and oscillation compared with the feedforward-feedback control strategy and strategy 1, and the system finally reached a steady state of 50°C. The results proved that the PID realized by the discrete differential can achieve higher control quality than the feedforward-feedback compound control. This algorithm realizes the filtering effect of high-frequency noise in the high-performance hot water supply system even without the feedforward filtering link. Therefore, the structure of the high-performance hot water supply system could be simplified by introducing the proposed PID controller.

Figure 11 shows the effect diagram of the high-performance hot water supply system with a PID controller to adjust the hot water setting temperature step by step without external interference. Curve 1 corresponds to the system effect when the PID controller's differential is an actual differential. Curve 2 corresponds to the system effect when the PID controller's differential link adopts the three-point discrete differentiator with an increasing sample step with $\Delta t = 1s$.

In **Figure 11**, the temperature is between 56.99°C and 57°C, and the time is from 400 seconds to 1000 seconds. The amplitude error is within 5%, and the system is in a stable state that the water temperature can meet the demand of people's life. Furthermore, the result indicates that the effect of the system with an actual

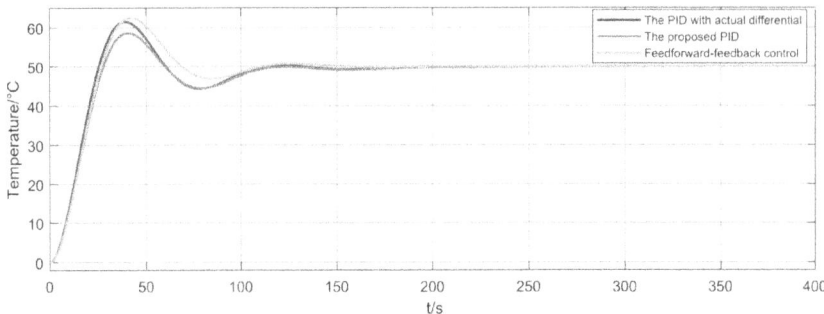

Figure 10.
Comparison of the results of the three methods.

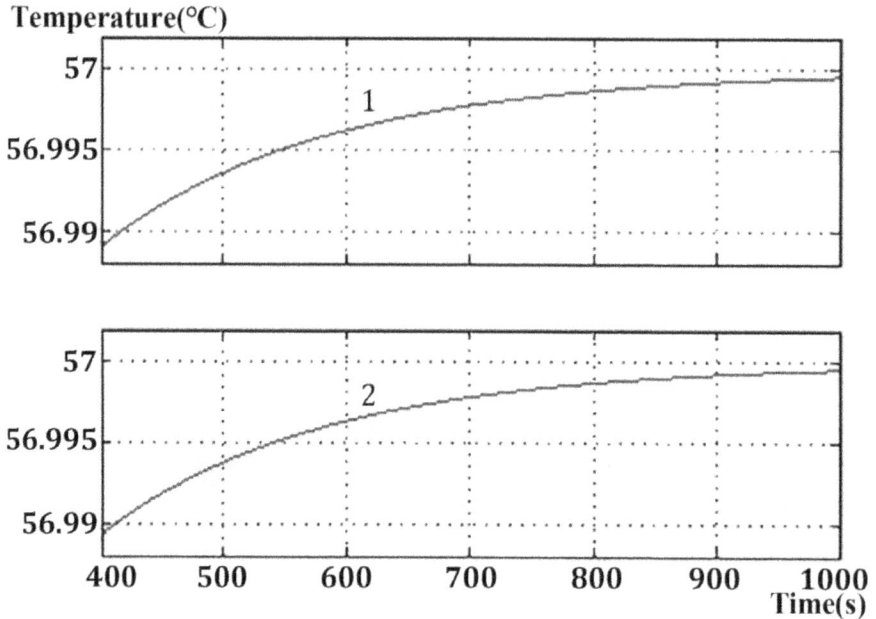

Figure 11.
Water temperature control effect with the proposed method and the actual differential.

differential in the PID controller is consistent with the system's effect while introducing the two-point differentiator filter with an increased sample step in the PID controller. It shows that the proposed algorithm can ensure that the system's transient response is not overshot.

Table 5 shows the standard deviation estimation of the input and output of the system under the different noise f, in which the PID controllers adopt the actual differential and the proposed differential.

The standard deviation estimation of the input and output of the continuous system (PID controller with actual differential) and the PID controller with the proposed differential is very similar. And when the four-point differentiator is adopted, the output standard deviation of the system is smaller than that of the system with actual differential. In addition, we have got the conclusion that the number of sampling points of the proposed method cannot exceed 4 in **Figures 3** and **4**. It shows that when the number of sampling points is 4, the differential error is the smallest and the control effect is the best.

The three-position relay with the dead zone characteristic can reduce the number of actuator actions and suppress the continuous small oscillations due to the quantization of the output of the PID controller and, at the same time, it simplifies the structure of the amplifier (heat exchanger) in the system.

However, the stability and control quality in such a nonlinear system depend on the relationship between $\pm a$ and $\pm B$. Next, we will introduce how to get the relationship.

The description function (harmonic linearization coefficient) of the relay with dead zone characteristics is shown in Eq. (19):

$$K(A) = \frac{\sqrt{1 - (a/A)^2} \cdot 4B}{\pi A}, (A \geq a) \tag{19}$$

Noise	Input	Actual differential	The proposed differentiator		
			2	3	4
Normal distribution Signal	0.24	6.65	6.77		
		6.92		7.07	
		7.11			6.94
White noise	0.77	7.2	7.16		
		8.31		8.14	
		9.64			9.15
Uniform distribution signal	0.44	6.8	6.87		
		7.31		7.36	
		7.86			7.55

Table 5.
The standard deviation of the output of the proposed new PID controller designed by the discrete differentiators with 2–4 sampling points in the high-performance hot water supply system.

In Eq. (19), A is the amplitude of input harmonic signal. a is the dead zone width $\pm a$ and B is relay response level $\pm B$.

In the dead zone relay, $K(A)$ would take the peak when $= \sqrt{2} \cdot a$. The peak of $K(A)$ is given in Eq. (20).

$$K(A) = 0.673 \cdot \frac{B}{a} \qquad (20)$$

The dead zone width a take ± 0.5 ($a = \pm 0.5$), we can get the relay response level B in Eq. (21) through the relation $K(A) \leq 1$ and (20):

$$B \leq 0.785 \qquad (21)$$

Table 6 lists the number of actions of the control device under white noise interference. PID controllers in the system respectively adopt the proposed differentiator and the actual differential, the system with the former one has smaller operation times of equipment. And as the number of sampling points in the algorithm increases, the number of actions of the control equipment also increases.

Filter	$W(s) = \frac{0.4}{(5s+1)}$			$W(s) = \frac{0.4}{(s+1)}$			$W(s) = \frac{0.4}{(0.5s+1)}$		
Number of the sampling points	2	3	4	2	3	4	2	3	4
m of control equipment 1	33	46	67	55	111	135	80	132	189
m of control equipment 2	28	29	52	32	76	100	50	81	136

m—action number of the control equipment in the high-performance hot water supply system under white noise.
control equipment 1—control equipment with the traditional PID.
control equipment 2—control equipment with the proposed PID designed by the discrete differentiators with 2–4 sampling points.

Table 6.
The action number of the control equipment which applied the PID controller designed by the discrete differentiators with 2–4 sampling points in the high-performance hot water supply system under white noise.

5. Conclusions

Aiming at the problem that differential link would introduce the high-frequency interference while improving the dynamic characteristics of the system, this paper proposes a numerical differential algorithm based on the equally-spaced Newton interpolation. Then, a discrete differentiator was constructed by and proposed the concept of "algorithm bandwidth" to ensure the accuracy of the effect of differential. Subsequently, the impact of the number of sampling points and the sampling step on the performance of the "discrete differentiator" was studied. It is concluded that the differentiator filter has the best differential accuracy and the strongest anti-noise ability for high-frequency noise when 2–4 sampling points and maximum allowable sample step are selected. Then, we designed a new PID controller based on a discrete differentiator. In order to verify the proposed PID controller's effect, some numerical simulations were given to verify its filtering effect. The results demonstrated that the disturbance introduced by high-frequency noises could be suppressed without extra filtering operation. Finally, we given an actual case of the high-performance hot water supply system, which adopted the cascade control scheme, and the temperature controller of the system applied the proposed PID controller. The results show that the PID controller designed by this algorithm has a feedforward filtering function even if the control system did not have a complex feedforward link or add an additional filter to suppress the high frequency noise in the system. At the same time, the introduction of the three-position relay effectively decreases the number of actuator actions. Compared with the traditional PID controller, the proposed PID controller with the discrete differentiator can achieve the effect of signal differentiation faster with a smaller mean square error, that is, it reaches the effect of "differentiation and filtering."

Acknowledgements

We want to thank to Zheng Liu for kind suggestions in the early version of this manuscript. The authors gratefully acknowledge the financial support of the Key Fund of Natural Science Foundation of Shaanxi Province (2019JLZ-06) and Key Project of Ministry of Industry and Information Technology of the People's Republic of China (2018-470).

Conflict of interest

We declare that we do not have any commercial or associative interest that represents a conflict of interest in connection with the work submitted. No potential conflict of interest was reported by the authors.

Author details

Biao Wang* and Shaojun Lin
The School of Electronics and Control Engineering, Chang'an University, Xi'an, China

*Address all correspondence to: wangbiao@chd.edu.cn

IntechOpen

References

[1] Beerens R, Bisoffi A, Zaccarian L, Heemels W, Nijmeijer H, van de Wouw N. Reset integral control for improved settling of PID-based motion systems with friction. Automatica. 2019; **107**:483-492. DOI: 10.1016/j. automatica.2019.06.017

[2] Yu H, Guan Z, Chen T, Yamamoto T. Design of data-driven PID controllers with adaptive updating rules. Automatica. 2020;**121**:109185. DOI: 10.1016/j.automatica.2020.109185

[3] Schuhmann T, Hofmann W, Werner R. Improving operational performance of active magnetic bearings using Kalman filter and state feedback control. IEEE Transactions on Industrial Electronics. 2011;**59**:821-829. DOI: 10.1109/TIE.2011.2161056

[4] Bruce LM, Li J. Wavelets for computationally efficient hyperspectral derivative analysis. IEEE Transactions on Geoscience and Remote Sensing. 2001; **39**:1540-1546. DOI: 10.1109/36.934085

[5] Poznyak AS, Yu W. Robust asymptotic neuro-observer with time delay term. International Journal of Robust and Nonlinear Control: IFAC-Affiliated Journal. 2000;**10**:535-559. DOI: 10.1109/ISIC.2000.882893

[6] Alwi H, Edwards C. An adaptive sliding mode differentiator for actuator oscillatory failure case reconstruction. Automatica. 2013;**49**:642-651. DOI: 10.1016/j.automatica.2012.11.042

[7] An H, Fidan B, Wu Q, Wang C, Cao X. Sliding mode differentiator based tracking control of uncertain nonlinear systems with application to hypersonic flight. Asian Journal of Control. 2019;**21**:143-155. DOI: 10.1002/asjc.1932

[8] He S, Wang J, Lin D. Composite guidance laws using higher order sliding mode differentiator and disturbance observer. Proceedings of the Institution of Mechanical Engineers, Part G: Journal of Aerospace Engineering. 2015;**229**: 2397-2415. DOI: 10.1177/09544100 15576365

[9] Kikuuwe R, Pasaribu R, Byun G. A first-order differentiator with first-order sliding mode filtering. IFAC-PapersOnLine. 2019;**52**:771-776. DOI: 10.1016/j.ifacol.2019.12.056

[10] Feng J, Wang W, Chen Y. An improved tracking-differentiator filter based on Taylor's formula. Optik. 2018; **158**:1026-1033. DOI: 10.1016/j.ijleo.2017. 12.198

[11] Wang G, Wang B, Zhao N, Xu D. A novel filtering method based on a nonlinear tracking differentiator for the speed measurement of direct-drive permanent magnet traction machines. Journal of Power Electronics. 2017;**17**: 358-367. DOI: 10.6113/JPE.2017.17.2.358

[12] Bertsias P, Psychalinos C. Differentiator based fractional-order high-pass filter designs. In: 2018 7th International Conference on Modern Circuits and Systems Technologies (MOCAST); 5-7 May 2018; Thessaloniki, Greece: IEEE; 2018. pp. 1-4. DOI: 10.1109/MOCAST43348.2018

[13] Hildebrand FB. Introduction to Numerical Analysis. Courier Corporation; 2nd ed. New York: Dover Publications; 1987. 669 p.

[14] Bazán FS, Bedin L. Filtered spectral differentiation method for numerical differentiation of periodic functions with application to heat flux estimation. Computational and Applied Mathematics. 2019;**38**:165. DOI: 10.1007/s40314-019-0968-4

[15] Chapra SC, Canale RP. Numerical Methods for Engineers.

Boston: McGraw-Hill Higher Education; 2010

[16] Schmitz G, Christiansen O. Gaussian process regression to accelerate geometry optimizations relying on numerical differentiation. The Journal of Chemical Physics. 2018; **148**:241704. DOI: 10.1063/1.5009347

[17] Zill D, Wright WS. Advanced Engineering Mathematics; 5th ed. US: Jones & Bartlett Learning; 2012. 1020 p.

[18] Chen B, Zhao Z, Li Z, Meng Z. Numerical differentiation by a Fourier extension method with super-order regularization. Applied Mathematics and Computation. 2018;**334**:1-10. DOI: 10.1016/j.amc.2018.04.005

[19] King JT, Murio DA. Numerical differentiation by finite-dimensional regularition. IMA Journal of Numerical Analysis. 1986;**6**:65-85. DOI: 10.1093/imanum/6.1.65

[20] Liu Y, Sun H, Yin X, Xin B. A new Mittag-Leffler function undetermined coefficient method and its applications to fractional homogeneous partial differential equations. Journal of Nonlinear Sciences and Applications. 2017;**10**:4515-4523. DOI: 10.22436/jnsa.010.08.43

[21] Abdulghafor R, Turaev S. Consensus of fractional nonlinear dynamics stochastic operators for multi-agent systems. Information Fusion. 2018;**44**: 1-21. DOI: 10.1016/j.inffus.2017.11.003

[22] Carnicer JM, Khiar Y, Peña JM. Optimal interval length for the collocation of the Newton interpolation basis. Numerical Algorithms. 2019;**82**: 895-908. DOI: 10.1007/s11075-018-0632-x

[23] Yang Y, Liang Y, Pan Q, Qin Y, Wang X. Gaussian-consensus filter for nonlinear systems with randomly delayed measurements in sensor networks. Information Fusion. 2016;**30**: 91-102. DOI: 10.1016/j.inffus.2015.12.003

[24] Chen Y, Qi G, Li Y, Sheng A. Diffusion Kalman filtering with multi-channel decoupled event-triggered strategy and its application to the optic-electric sensor network. Information Fusion. 2017;**36**:233-242. DOI: 10.1016/j.inffus.2016.12.004

[25] Manju B, Sneha M. ECG denoising using Wiener Filter and Kalman Filter. Procedia Computer Science. 2020;**171**: 273-281. DOI: 10.1016/j.procs.2020.04.029

Section 3

Smart City Essentials

Traffic State Prediction and Traffic Control Strategy for Intelligent Transportation Systems

Shangbo Wang

Abstract

The recent development of V2V (Vehicle-to-Vehicle), V2I (Vehicle-to-Infrastructure), V2X (Vehicle-to-Everything) and vehicle automation technologies have enabled the concept of Connected and Automated Vehicles (CAVs) to be tested and explored in practice. Traffic state prediction and control are two key modules for CAV systems. Traffic state prediction is important for CAVs because adaptive decisions, control strategies such as adjustment of traffic signals, turning left or right, stopping or accelerating and decision-making of vehicle motion rely on the completeness and accuracy of traffic data. For a given traffic state and input action, the future traffic states can be predicted via data-driven approaches such as deep learning models. RL (Reinforcement Learning) - based approaches gain the most popularity in developing optimum control and decision-making strategies because they can maximize the long-term award in a complex system via interaction with the environment. However, RL technique still has some drawbacks such as a slow convergence rate for high-dimensional states, etc., which need to be overcome in future research. This chapter aims to provide a comprehensive survey of the state-of-the-art solutions for traffic state prediction and traffic control strategies.

Keywords: traffic state prediction, traffic control, deep learning, V2V, V2I, V2X, CAV, RL

1. Introduction

Connected and Automated Vehicles (CAVs) are nowadays the area of extensive research and there are premises to suspect that the introduction CAVs may revolutionize the whole transportation area [1]. There is no lack of predictions stating that CAVs will solve many of the current problems experienced on roads today, such as congestion, traffic accidents and lost time [2].

Traffic state prediction and traffic control are two key modules in transportation systems with CAVs [3]. Traffic states such as flow, speed, congestion, etc., plays vital roles in traffic management, public service and traffic control [4]. By predicting the evolution of traffic state timely and accurately, decision-maker and traffic controller can make effective policy and control input to avoid traffic congestion ahead of time and thus ITS (Intelligent Transportation Systems), advanced traffic management systems and traveler information systems rely on real-time traffic state prediction. Traffic control can be divided into a decision-making module and a vehicle control module. The former is used to optimize the mobility, safety and energy consumption by using the vehicle trajectory prediction results to calculate vehicle platoon sizes, speed, flow, density, traffic merging, diverging flow and traffic signals, while the latter is used for vehicle path control, vehicle fleet control and steering wheel, throttle, brake, and other actuator control by using onboard units based on the control commands [3]. How to timely and accurately predict the future traffic state and deliver an effective traffic control strategy are fundamental issues in ITS.

Traffic state prediction approaches can be broadly divided into two parts: parametric and non-parametric approaches [5]. Parametric approaches utilize parametric models that capture all the information about its predictions within a finite set of parameters. The popular techniques in parametric approaches include ARIMA (Autoregressive Integrated Moving Average) [6–9], linear regression [10] and Kalman Filter (KF) based method [11], which are linear models and able to have high accuracy with linear characteristics of traffic data. ARIMA model is based on the assumption that the future data will resemble the past and widely used in time series analysis, which can be made to be stationary by differencing. It can be specified three values that represent the order of autoregressive (p), the degree of differencing (d) and the order of moving average (q). The model order can be selected by Akaike's Information Criterion (AIC) combined with the likelihood of the historical data while the model parameters can be estimated by maximizing the log likelihood function. The extension of the ARIMA time series model into the spatial domain results in the STARIMA (Space–Time Autoregressive Integrated Moving Average) model, which can deliver more accurate prediction results in traffic prediction because of spatio-temporal correlation of traffic data. To capture the spatial correlation, the STARIMA model adds the spatial matrix comprising spatial adjacency and weight structure, and the number of spatial lags for STAR and STMA models. The drawback of the ARIMA model for traffic prediction is the strong assumption that traffic data can become stationary by differencing, which is difficult to be fulfilled because of the non-stationary characteristics of traffic data. By assuming the linear relationship between the input variables and the single output variable, the linear regression model aims to estimate the regression coefficients by using the historical traffic data. KF based methods allow a unified approach for the prediction of all processes that can be given a state space representation. Although, EKF (Extended Kalman Filtering) can be used to deal with the non-linearity of traffic data, it is difficult to have an accurate approximation of most non-linear functions and thus it can lead to relatively large error.

Non-parametric models such as DL (Deep Learning) outperform parametric models because of stochastic, indeterministic, non-linear and multidimensional characteristics of traffic data [5]. DL is a subset of machine learning (ML) which is based on the concept of deep neural network (DNN) and it has been widely used for data classification, natural language processing (NLP) and object recognition [5]. The most popular DL models used for traffic state prediction includes Convolution Neural Network (CNN) [12–14], Deep Belief Network (DBN) [15, 16], Recurrent Neural Network (RNN) [17–19] and Autoencoder (AE) [20] etc. CNN is useful for traffic prediction because of the two-dimensional characteristics of traffic data and its ability to extract the spatial feature. CNN is only connected to a smaller subset of input and thus decreases the computational complexity of the training process. DBN is a stacking of multiple RBMs (Restricted Boltzmann Machines), which can be used to estimate the probability distribution of the input traffic data. LSTM is the special type of RNN, which can capture the temporal feature of traffic data, and LSTM can overcome the gradient vanishing problem caused by the standard RNN.

Traffic control strategies can be generally divided into classical methods and learning-based methods. Classical methods develop traffic controller based on control theory or optimization-based techniques, which include dynamic traffic assignment based nonlinear controller [21], standard proportional-integral (PI) controller [22, 23], robust PI controller [24], model-based predictive control (MPC) [25, 26], linear quadratic controller [27], mixed-integer non-linear programming (MINLP) [28], multi-objective optimization based decision-making model [29]. Learning-based methods refer to the utilization of artificial intelligence technologies to achieve decision-making and control for CAVs, which can be further divided into three categories: statistic learning-based method, deep learning-based (DL) method and reinforcement learning-based (RL) method. The RL-based method is currently one of the most commonly used learning-based techniques for traffic control and decision-making because RL can solve complex control problems by using the Markov decision process (MDP) to describe the interaction states of agent and environment [4]. The most popular RL-based methods include Q-learning for adaptive traffic signal control [30, 31], multi-agent RL approaches [32–35], Nash Q-learning strategy [36]. Many other RL-based approaches are also available in the literature. Q-learning based traffic signal control aims to minimize the average accumulated travel time by greedily selecting action at each iteration. Multi-agent RL approaches are more popularly used in network signal optimization and can be generally divided into centralized RL and decentralized RL, while the former considers the whole system as a single agent and the latter distributes the global control to each agent. Nash Q-learning strategy is a decentralized multi-agent RL strategy, which performs iterated updates based on assuming Nash equilibrium behavior over the current Q-values. It can be shown that traffic signal control using the Nash Q-learning strategy can converge to at least one Nash equilibrium for stationary control policies. However, Nash Q-learning is unable to achieve the Pareto Optimality without consideration of cooperation among different agents.

This chapter provides a comprehensive survey about state-of-the-art traffic state prediction and traffic control techniques. It is organized as follows: In Section 2, we firstly introduce the fundamental structure and main characteristics of two important DL models: CNN and LSTM (Long Short-Term Memory), as well as their advantages in traffic state prediction, then we introduce how to realize hybrid traffic state prediction by combining two models to achieve better accuracy. In Section 3, we detail RL fundamentals and introduce how it can be applied in traffic control and decision-making. We focus on multi-agent RL approaches. Pros and cons are discussed. Section 4 gives the summary of this chapter.

2. DL-based traffic state prediction approaches

In this section, we first briefly overview the machine learning and deep learning concept. Then, we focus on introducing the architectures of two DL models: CNN and LSTM, which show good performance in processing high-dimensional and temporal correlated data. Finally, a hybrid model of CNN and LSTM is described and the research potential is about how to improve the prediction accuracy by incorporating spatio-temporal correlation.

2.1 Overview of deep learning

ML approaches are broadly classified into two categories, i.e., Supervised Learning and Unsupervised Learning [5]. Supervised Learning requires input data to be clearly labeled. It involves a function $y = f(x)$ that maps input x to output y [5]. It aims to perform two tasks: regression and classification. In contrast to supervised learning, unsupervised learning aims to perform data clustering by extracting the data pattern. Some popularly used supervised learning approaches mainly include Random Forest (RF), Support Vector Machine (SVM), Bayesian methods, Artificial Neural Network (ANN), etc., whereas unsupervised learning approaches mainly include Autoencoder, Principal Component Analysis (PCA), Deep Belief Network (DBN), etc. To perform prediction tasks by supervised learning approaches, a model needs to be trained firstly by a training dataset with a certain amount of samples. Then, new predictions can be obtained by inputting the feature vector of new samples to the trained model. Cross-validation is usually used to validate the prediction performance by performing the following procedures for each k-th fold: (i) the whole dataset is divided into K folds; (ii) a model is trained by using $K - 1$ folds as training data; (iii) the resulting model is validated by the k-th fold.

DL is a branch of ML which aims to construct a computational model with multiple processing layers to support high-level data abstraction. It can automatically extract the feature from data, without any human interference to explore hidden data relationships among different attributes of the dataset [37]. Concepts of DL are inspired by the thinking process of the human brain. Hence, the majority of DL architectures are using the framework of Artificial Neural Network (ANN), which consists of input, hidden and output layers with nonlinear computational elements (neurons and processing units). The network depth (the number of layers) can be adjusted according to the feature dimensions and complexity of the data. The number of neurons at the input layer is equal to the number of independent variables, while the number of neurons at the output layer is equal to the number of dependent variables, which can be single or multiple. Neurons of two successive layers are connected by weights which are updated while training the model. The neurons at each layer receive the output from the previous layer, which is generated by a weighted summation over inputs and then passed to an activation function (**Figure 1**).

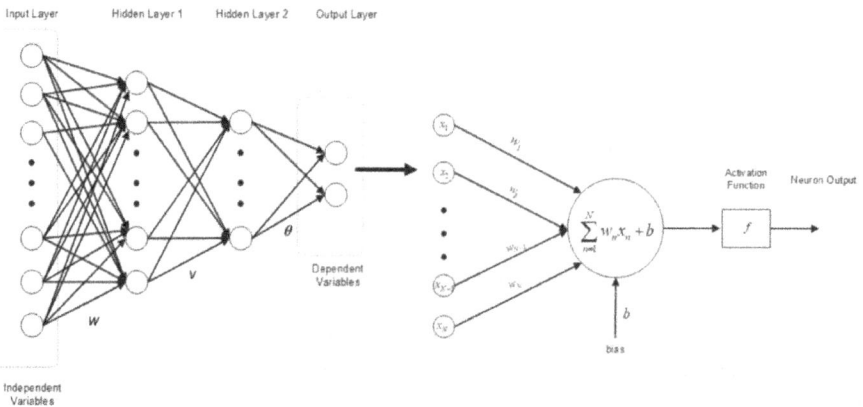

Figure 1.
*(left) ANN with one input layer, two hidden layers and one output layer, **W**, **V**, θ: Weighting matrices between IL and HL1, HL1 and HL2, HL2 and OL; (right) an illustration of an output generation.*

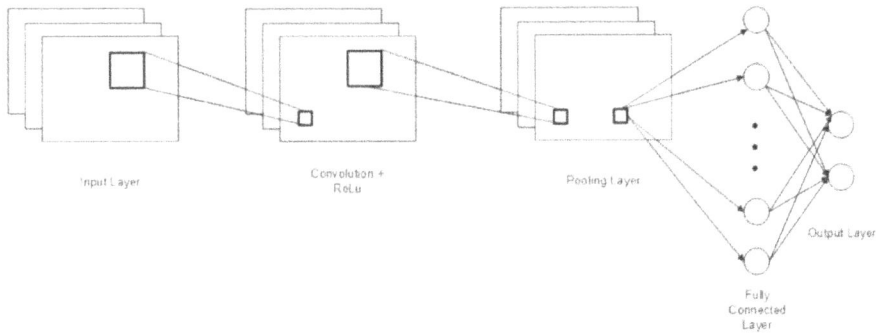

Figure 2.
CNN structure.

Let us take the four-layer ANN in **Figure 2** for example. During the training process, the value of k-th neuron at the output layer can be calculated by

$$o_k = f\left(\sum_{j=1}^{J} \theta_{k,j} z_j + b_k\right) \tag{1}$$

$$z_j = f\left(\sum_{m=1}^{M} v_{j,m} y_m + b_j\right) \tag{2}$$

$$y_m = f\left(\sum_{n=1}^{N} w_{m,n} x_n + b_m\right) \tag{3}$$

where f is the activation function, x_n is the independent variable at the n-th neurons of the input layer, b_k, b_j and b_m are bias values, N, M and J are the numbers of neurons at the input layer, hidden layer 1 and hidden layer 2, respectively. Then, the unknown parameters W, V, θ are adjusted to minimize loss function such as MSE (Mean Squared Error) by back propagation algorithms [38, 39].

2.2 Fundamental structure of CNN and LSTM

In this section, we examine two popular DL architectures: CNN and LSTM, which are used popularly for multidimensional and time sequential dataset. CNNs have been extensively applied in various fields, including traffic flow prediction [14, 40, 41], computer vision [42], Face Recognition [43], etc., while LSTMs are special kinds of RNNs, which are mainly applied in the area of temporal data processing, such as traffic state prediction [34, 44], speech processing [45] and NLP (Natural Language Processing) contexts [46], etc.

The significant difference between fully connected ANN and CNN is that CNN neurons are only connected to a smaller subset of input which decreases the total parameters in the network [47]. CNNs have the ability to extract important and distinctive features from multidimensional by making use of filtering operations. A commonly used type of CNNs, which is similar to multi-layer perception (MLP), consists of numerous convolution layers preceding pooling layers and fully connected layers. CNN structure is illustrated in **Figure 2**, where it consists of the input layer, convolution layer, pooling layer and fully connected layer. Convolution layer outputs higher abstraction of the feature. Each convolution layer uses several filters, which are designed to have a distinct set of weights. Filters used by the convolution layer have the smaller dimensions compared to the data size. In the training phase, filter weights are automatically determined according to an assigned task. The filters of each convolution layer are applied through the input layer by computing the sum of the product of input and filter, leading to a feature map of each filter. Each feature map detects a distinct high-level feature which is then processed by a pooling layer and a fully connected layer. ReLU activation function is applied to remove all negative values in the feature map.

The benefits of CNNs over other statistical learning methods and DL methods are listed followings [48]:

1. CNNs have the weight sharing feature, which reduces the number of trainable network parameters and in turn helps the network speed up the training process and avoid overfitting.

2. Concurrently learning the feature extraction layers and the classification layer causes the model output to be both highly organized and highly reliant on the extracted features.

3. Large-scale network implementation is much easier with CNN than with other neural networks.

CNN and other kinds of ANNs such has MLP are not designed for sequences and time series data because they do not have memory element. In such cases, RNN can deliver more accurate results. RNNs are widely used in traffic state prediction because traffic data has spatiotemporal characteristics, which cannot be captured by CNN or other kinds of ANNs. RNN structure is illustrated in **Figure 3**, where RNNs involve an internal memory element that memorizes the previous output. The current output h_t is not only based upon present input x_t, but also on previous output h_{t-1}, which can be expressed by

$$h_t = f_h(W_i x_t + W_r h_{t-1} + b_h) \tag{4}$$

$$y_t = f_t(W_o h_t + b_y) \tag{5}$$

where W_i, W_r and W_o are respectively the weighting matrices for the current input vector x_t, previous output vector h_{t-1} and current output vector h_t, b_h and b_y are the bias vectors, f_h and f_t are the activation functions.

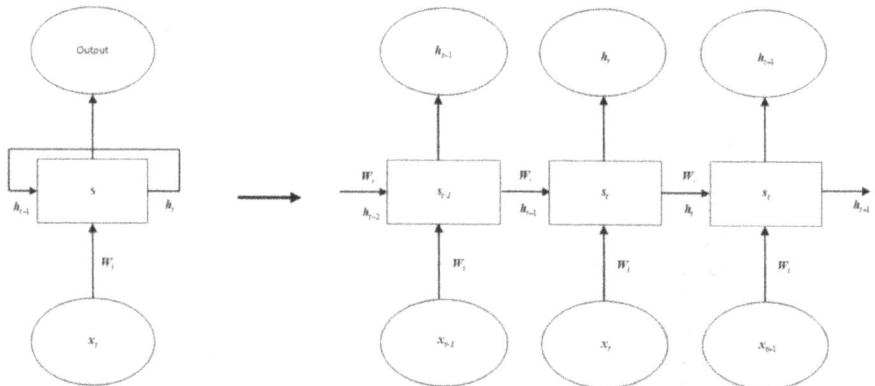

Figure 3.
RNN structure.

LSTM is firstly proposed in [49] to overcome the gradient vanishing problems generated by other RNNs. A typical LSTM network consists of an input layer, a recursive hidden layer and an output layer. In the recursive hidden layer, each neuron is made up of four structures: a forget gate, an input gate, an output gate and a memory block. The state of the memory cell reflects the features of the input, while the three gates can read, update and delete features stored in the cell. The LSTM structure is illustrated in **Figure 4**.

The past information carried by the cell state c_{t-1} can be regulated by the current input state x_t, the previous output h_{t-1} and the gate σ, which is usually composed of a sigmoid neural network layer and a pointwise multiplication operation. From **Figure 4**, the forget gate f_t, input gate i_t, cell state c_t, output gate o_t, and hidden state h_t at t-th time instant can be expressed by

$$f_t = \sigma\left(W_f[h_{t-1}, x_t] + b_f\right)$$
$$i_t = \sigma\left(W_i[h_{t-1}, x_t] + b_i\right)$$
$$c_t = f_t \odot c_{t-1} + i_t \odot tanh\left(W_c[h_{t-1}, x_t] + b_c\right) \qquad (6)$$
$$o_t = \sigma\left(W_o[h_{t-1}, x_t] + b_o\right)$$
$$h_t = o_t \odot tanh\left(c_t\right)$$

where W and b are respectively weighting matrix and bias vector and . denotes the subscripts including f, i, c and o, $\odot$ is point-wise multiplication. The forget gate decides which information needs attention and which can be ignored. The information from the current state x_t and hidden state h_{t-1} are passed through the sigmoid function. Sigmoid generated values between 0 and 1. It concludes whether the part of the old output is necessary. The input gate updates the cell state by the following operations: first, values between 0 and 1 are generated by passing the current state x_t and previous hidden state h_{t-1} into the second sigmoid function. Then, the same information of the hidden state and current state will be passed through the tanh function to generate values regulated by the first operation. Finally, the current cell state c_t is updated by weighted summation of the generated values and past cell state c_{t-1}. The output gate determines the value of next hidden state. First, the values of the current state x_t and previous hidden state h_{t-1} are passed into the third sigmoid function. Then, the new cell state generated from the cell state is passed through the tanh function. Based on the final value, the network decides which information the hidden state should carry.

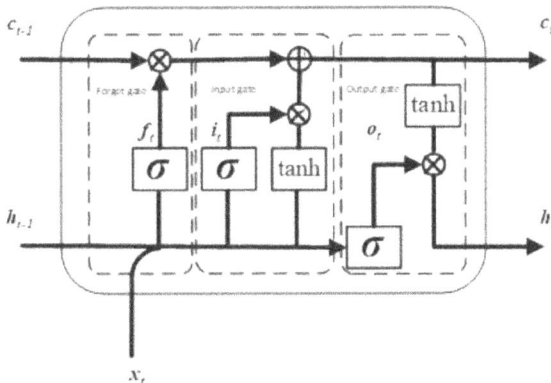

Figure 4.
LSTM structure.

Generally, LSTM can address the vanishing gradient problem that makes network training difficult for a long-sequence temporal data. The long-term dependencies in the data can be learned to improve the prediction accuracy.

2.3 Traffic state prediction with hybrid DL models

Although, CNN and LSTM have advantages in dealing with traffic data with spatiotemporal dependencies, due to the complex and non-linear models of traffic data, it is hard to predict accurate results by using a single model [5]. Some litera-ture proposed that prediction accuracy can be improved by hybrid modeling such as combining CNN and LSTM [50–53].

The spatial and temporal features can be fully extracted by hybrid models, where CNN in this model is used to capture spatial features of traffic data whereas LSTM is used to extract temporal features. Suppose that we have traffic state data of K locations $\{s_{i=1}^{K}\}$ in $(t - N, t - 1)$ are used as inputs to predict the traffic states at $(t, t + 1, \cdots, t + h)$. The real-time measured data can be arranged into a matrix:

$$S = \begin{bmatrix} s_1 \\ s_2 \\ \vdots \\ s_K \end{bmatrix} = \begin{bmatrix} s_{1,t-N} & s_{1,t-N+1} & \cdots & s_{1,t-1} \\ s_{2,t-N} & s_{2,t-N+1} & \cdots & s_{2,t-1} \\ \vdots & \vdots & \ddots & \vdots \\ s_{K,t-N} & s_{K,t-N+1} & \cdots & s_{K,t-1} \end{bmatrix} \tag{7}$$

Note that $s_{k,t}$ in the matrix can also be a vector which may include flow, speed, position, etc. Traffic state such as traffic flow depicts spatio-temporal characteris-tics, that is, the traffic state on each location at a certain time instant depends on that of neighboring locations at current or different time instant.

There are mainly two hybridization manners: the first one is to extract spatio-temporal features by concatenating CNN and LSTM, that is, each column of S is firstly input into a CNN model to obtain the high-level spatial feature map, which is then input into a LSTM.

model to capture the temporal features; the second one is to parallelize CNN and LSTM modeling process by considering the extracted spatial and temporal features are of the same importance, that is, the same traffic state data is input into two models, the final prediction is obtained by passing the output of two models through a FC (Fully Connected) layer. The structure of the two hybridizations is illustrated as follows (**Figure 5**).

For concatenated hybrid models, the real-time measured data matrix S is firstly parallelized in the time domain; then, a one-dimensional CNN is used to capture

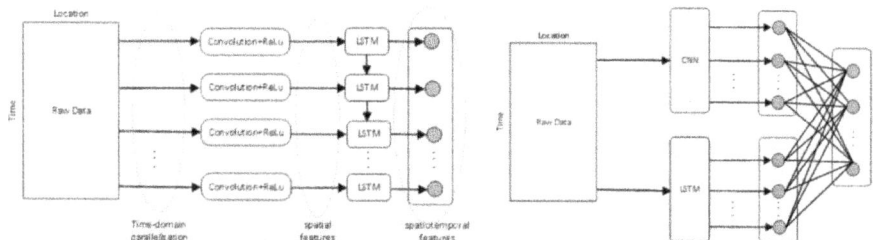

Figure 5.
(left) concatenated hybrid model; (right) parallelized hybrid model.

high-level spatial features for each channel; finally, the high-level spatial feature map is input into LSTM models to generate the final prediction.

The high-level spatial feature map output by the one-dimensional CNN can be expressed by

$$
X = \begin{bmatrix} x_1 \\ x_2 \\ \vdots \\ x_L \end{bmatrix} = \begin{bmatrix} x_{1,t-N} & x_{1,t-N+1} & \cdots & x_{1,t-1} \\ x_{2,t-N} & x_{2,t-N+1} & \cdots & x_{2,t-1} \\ \vdots & \vdots & \ddots & \vdots \\ x_{L,t-N} & x_{L,t-N+1} & \cdots & x_{L,t-1} \end{bmatrix} \tag{8}
$$

where $x_{l,t}$ is the l-th high-level feature at the t-th time instant and can be obtained as follows

$$
x_{l,t} = f_l(w_t \otimes S_t + b_t) \tag{9}
$$

where w_t is the one-dimensional filter with $K - L + 1$ coefficients, $\otimes$ is the convolution operation, S_t is the t-th column of S, b_t is the bias, f_l denotes the l-th activation function. For simplicity, only one convolution layer in Eq. (9) is displayed. In practice, multiple convolutions and pooling layers can be used to satisfy the demand according to the data size.

To extract the temporal features, the high-level spatial feature vector for single or multiple time instants will be selected for the input of each LSTM, which is denoted as

$$
F = [F_0 \ F_1 \ \cdots \ F_{N-1}] \tag{10}
$$

where F_n is the high-level spatial feature map for the n-th LSTM network denoted as

$$
F_n = \begin{bmatrix} x_{1,t+n-N} & x_{1,t+n-N+1} & \cdots & x_{1,t+n-N+M-1} \\ x_{2,t+n-N} & x_{2,t+n-N+1} & \cdots & x_{2,t+n-N+M-1} \\ \vdots & \vdots & \ddots & \vdots \\ x_{L,t+n-N} & x_{1,t+n-N+1} & \cdots & x_{L,t+n-N+M-1} \end{bmatrix} \tag{11}
$$

where M is the adjustable input window size. The spatio-temporal features output by LSTM are denoted as $H = [H_0 \ H_1 \ \cdots \ H_{N-1}]$, H_n is a $K \times T$ matrix, where T is the adjustable output window size with $M + T \le N$. Combining with **Figure 4** and Eq. (6), the generated spatio-temporal features are iteratively determined by

$$
\begin{aligned}
\mathbf{f_n} &= \sigma\big(\mathbf{W}_{f,F}\mathrm{vec}(\mathbf{F}_n) + \mathbf{W}_{f,H}\mathrm{vec}(\mathbf{H}_{n-1}) + \mathbf{b_f}\big) \\
\mathbf{i_n} &= \sigma\big(\mathbf{W}_{i,F}\mathrm{vec}(\mathbf{F}_n) + \mathbf{W}_{i,H}\mathrm{vec}(\mathbf{H}_{n-1}) + \mathbf{b_i}\big) \\
\mathbf{c_n} &= \mathbf{f_n} \odot \mathbf{c_{n-1}} + \mathbf{i_n} \odot \tanh\big(\mathbf{W}_{c,F}\mathrm{vec}(\mathbf{F}_n) + \mathbf{W}_{c,H}\mathrm{vec}(\mathbf{H}_{n-1}) + \mathbf{b_c}\big) \\
\mathbf{o_n} &= \sigma\big(\mathbf{W}_{o,F}\mathrm{vec}(\mathbf{F}_n) + \mathbf{W}_{o,H}\mathrm{vec}(\mathbf{H}_{n-1}) + \mathbf{b_o}\big) \\
\mathbf{H}_n &= \mathbf{o_n} \odot \tanh(\mathbf{c_n})
\end{aligned} \tag{12}
$$

where $\mathbf{W}_{*,F}$ and $\mathbf{W}_{*,H}$ are respectively the weighting matrices for the current input high-level spatial feature matrix and previous spatio-temporal feature matrix, vec() is used for vectorization due to different size of $\mathbf{F}_n$ and $\mathbf{H}_{n-1}$, σ and *tanh* are respectively the sigmoid function and hyperbolic function with vector input.

Concatenated hybrid models utilize a one-dimensional CNN to obtain a smaller range of spatial features, in addition, they do not contain a fully connected layer at the output of LSTM models, and thus concatenated hybrid models are with low learning complexity. However, the temporal features delivered by LSTM have a strong correlation with the spatial features output by CNN, which needs some special assumptions about the raw data.

For parallelized hybrid models, the historical data matrix S need not be parallelized in the time domain, rather it is input into a CNN and LSTM simultaneously to extract the high-level spatial and temporal feature map independently. Then, the final prediction is generated by merging the high-level spatial and temporal features via a fully-connected layer. The high-level spatial feature map can be obtained by filtering S via two-dimensional CNNs. Suppose that we utilize a CNN with L convolution layers, the spatial filter for the l-th layer is denoted as $w_{i,j}^l, i = 0, 1, \cdots, I - 1; j = 0, 1, \cdots, J - 1$, where I and J are the size of the spatial filter. Given the historical data matrix S in (7), the output of the l-th layer of the n-th CNN is obtained by

$$
o_{i,j,n}^{0,c} = s_{i,j+n}
$$

$$
o_{i,j,n}^{l,c} = \sigma \left(\sum_{i'=0}^{I-1} \sum_{j'=0}^{J-1} w_{i',j'}^l o_{i+i',j+j',n}^{l-1,c} + b_{i,j}^l \right) \tag{13}
$$

$$
i = 0, 1, \cdots, I - 1; j = 0, 1, \cdots, J - 1, n = 0, 1, \cdots, N - 1
$$

where $o_{i,j,n}^{l,c}$ represents the (i,j)-th output of the l-th layer of the n-th CNN, σ is the activation function and $b_{i,j}^l$ is the (i,j)-th bias of the l-th layer of CNN.

A LSTM is utilized to obtain the high-level temporal feature map. The output of the n-th LSTM can be obtained by Eq. (12) with the input of S_n, which can be denoted as

$$
S_n = \begin{bmatrix} s_{1,t-N+n} & s_{1,t-N+n+1} & \cdots & s_{1,t-N+n+M-1} \\ s_{2,t-N+n} & s_{2,t-N+n+1} & \cdots & s_{2,t-N+n+M-1} \\ \vdots & \vdots & \ddots & \vdots \\ s_{K,t-N+n} & s_{K,t-N+n+1} & \cdots & s_{K,t-N+n+M-1} \end{bmatrix} \tag{14}
$$

$$
o_n^L = \text{vec}(H_n) = \begin{bmatrix} o_{0,n}^L & o_{1,n}^L & \cdots & o_{KT-1,n}^L \end{bmatrix}^T
$$

$$
n = 0, 1, \cdots, N - 1
$$

By posing a fully connected layer to the output of the L-th CNN layer $o^{L,c}$ and the output of the n-th LSTM o_n^L, the n-th final prediction can be obtained by

$$
o_n^F = \sigma \left(W^F \left[o_n^{L,c,T}, o_n^{L,T} \right]^T + b^F \right) = \begin{bmatrix} o_{0,n}^F & o_{1,n}^F & \cdots & o_{KT-1,n}^F \end{bmatrix}^T \tag{15}
$$

where W^F and b^F are the weighting matrix and bias vector for the fully connected layer, respectively.

In parallelized hybrid models, the spatial and temporal feature maps are considered to be of the same importance, and thus are extracted independently. The fully connected layer merges the output of CNN and LSTM without any special assumptions about the high-level spatial and temporal features.

Traffic state has strong periodic features because people get used to repeating some similar or same behaviors on the same time period of different days or the same day of different weeks, e.g., most people routinely go to work in the morning and go home in the evening during the peak hour [53]; most people routinely go for shopping on weekends rather than weekdays, etc. The periodic features can be used as supplementary information to predict the future traffic state. For the short-term traffic state prediction, the real-time data only contains the data before the prediction time instant, but the historical data on previous days or weeks contain the full data of that period, that means, traffic state information after the inspected time instant on previous days or weeks can be utilized to get the prediction about that on the inspected time instant. Suppose we use parallelized hybrid models, the complete prediction structure should contain CNN and LSTM for the real-time data, CNN and bidirectional LSTM for the historical data, which are connected by using a fully connected layer.

The bidirectional LSTM is composed of two independent forward and backward LSTMs, whose inputs are the time series before and after the inspected time instant. The final prediction of bidirectional LSTM is obtained by concatenating the forward and backward LSTMs. The structure of bidirectional LSTM is depicted in **Figure 6**.

Suppose that additionally, we have historical traffic state data S_d on the d-th day, which is denoted as

$$
\begin{aligned}
S_d &= \begin{bmatrix} s_{t-N,d} & \cdots & s_{t-1,d} & \cdots & s_{t+N,d} \end{bmatrix} \\
&= \begin{bmatrix}
s_{1,t-N,d} & s_{1,t-N+1,d} & \cdots & s_{1,t+N,d} \\
s_{2,t-N,d} & s_{2,t-N,d} & \cdots & s_{2,t+N,d} \\
\vdots & \vdots & \ddots & \vdots \\
s_{K,t-N,d} & s_{K,t-N,d} & \cdots & s_{K,t+N,d}
\end{bmatrix}
\end{aligned}
\tag{16}
$$

$$
d = 1, 2, \cdots, D
$$

where D is the number of previous days. We assume each previous day has data at $2N + 1$ time instants available for prediction. The input and output of the forward LSTM are given by Eqs. (14) and (15), and the input of the backward LSTM is given by

$$
S_{n,d,BL} = \begin{bmatrix}
s_{1,t+N-n-M+1,d} & s_{1,t+N-n-M+2,d} & \cdots & s_{1,t+N-n,d} \\
s_{2,t+N-n-M+1,d} & s_{2,t+N-n-M+2,d} & \cdots & s_{2,t+N-n,d} \\
\vdots & \vdots & \ddots & \vdots \\
s_{K,t+N-n-M+1,d} & s_{K,t+N-n-M+2,d} & \cdots & s_{K,t+N-n,d}
\end{bmatrix}
\tag{17}
$$

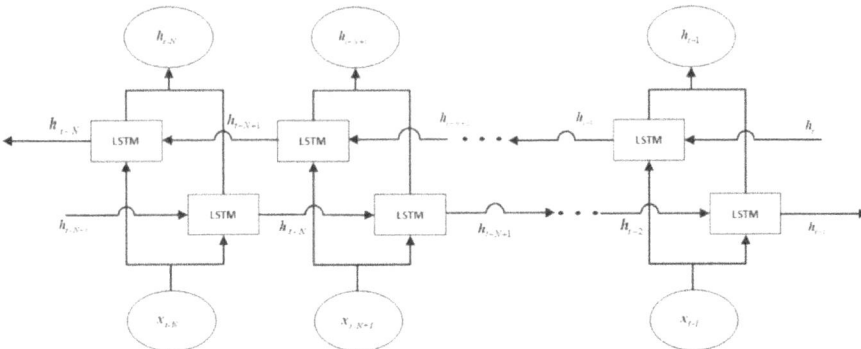

Figure 6.
Bidirectional LSTM structure.

Using Eq. (12), the output of the n-th backward LSTM can be obtained by

$$o_n^{BL} = \text{vec}(H_n^{BL}) = \begin{bmatrix} o_{0,n}^{BL} & o_{1,n}^{BL} & \cdots & o_{KT-1,n}^{BL} \end{bmatrix}^T \tag{18}$$

Then, the n-th temporal feature can be obtained by concatenating o_n^L and o_n^{BL}.

3. Reinforcement learning based traffic signal control

An accurate and efficient traffic state prediction can provide continuous and precise traffic status and vehicle states based on past information. How to utilize the current and predicted traffic states to make a real-time optimum decision is the main task of the traffic signal control module in ITS. The objectives of traffic signal control include minimizing the average waiting time at multiple intersections, reducing traffic congestion and maximizing network capacity. There exist real-time linear feedback control approaches and MPC (Model-based Predictive Control) that are specifically designed for traffic signal control systems to achieve the targets. The drawback of linear feedback control techniques that have been tried is that the system should always remain in the linear region at all times for the controller. Although, MPC has some advantages such as imposing constraints, the main shortcoming is it needs an accurate dynamic model, which is difficult to be obtained for traffic control systems. Data-driven approaches such as DRL (Deep Reinforcement Learning) based traffic control techniques are widely presented for ITS in recent years because RL can solve complex control problems and deep learning helps to approximate highly nonlinear functions from the complex datasets. In this section, we firstly briefly review the fundamental principles of RL. Then, we focus on multi-agent DRL based traffic signal control techniques such as decentralized multi-agent advantage actor-critic, which can converge to the local optimum and overcome the scalability issue by considering the non-stationarity of MDP transition caused by policy update of the neighborhood; and Nash Q-learning strategy, which can converge to Nash equilibrium by only considering the competition among agents.

3.1 Overview of reinforcement learning

Reinforcement Learning (RL) is a promising data-driven approach for decision-making and control in complex dynamic systems. RL methodology formally comes from a Markov Decision Process (MDP), which is a general mathematical framework sequential decision-making algorithms, and consists of five elements [54]:

- A set of states S, which contains all possible states the system can be in;

- A set of actions A, which contains all possible actions the system can respond;

- Transition probability $T(s_{t+1}|s_t, a_t)$, which maps a state-action pair for each time t to a distribution of next state s_{t+1};

- Reward function $r(s_t, a_t, s_{t+1})$ which gives the instantaneous reward for taking action a_t from the state s_t when transitioning to the next state s_{t+1};

- The discount factor γ between 0 and 1 for future rewards.

RL aims to maximize a numerically defined reward by interacting with the environment to learn how to behave in an environment without any prior knowledge by learning. In traffic signal control systems, RL is used to find the best control policy π^* that maximizes the expected cumulative reward $E[R_t|s, \pi]$ for each state s and cumulative discounted reward

$$R_t = \sum_{i=0}^{T-1} \gamma^i r_{t+i} \tag{19}$$

where γ is the discount factor that reflects the importance of future rewards, r_t is the t-th instantaneous reward. Choosing a larger γ means that the agent's actions have a higher dependency on future reward, whereas a smaller γ results in actions that mostly care about r_t.

RL generally can be classified into model-based RL which knows or learns the transition model from state s_t to s_{t+1}, and model-free RL which explores the environment without knowing or learning a transition model. Model-based RL emphasizes planning, that is, agents can keep track of all the routes in future time instants by predicting the next state and reward. Model-free RL can estimate the optimal policy without using or estimating the dynamics of the environments. In practice, model-free RL either estimates a value function or the policy by interacting with the environment and observing the responses. Model-free RL can be classified into value-based RL and policy-based RL. Value-based RL is mostly used for the cases that control problems have discrete state-action space. Q-learning and SARSA are two main value-based RL techniques, where the values of state-action pairs (Q-value) are stored in a Q-table, and are learned via the recursive nature of Bellman equations utilizing the Markov property [54]:

$$Q^\pi(s_t, a_t) = E_\pi[r_t + \gamma Q^\pi(s_{t+1}, \pi(s_{t+1}))] \tag{20}$$

where $\pi(a|s)$ is the probability of the action a is selected by given s, Q is the expected accumulative reward given by an action a and a state s

$$Q^\pi(s, a) = E[R_t|s, a, \pi] \tag{21}$$

The stochasticity in Eq. (21) comes from the control policy π and the transition probability from s_t to s_{t+1}. Value-based RL updates the Q^π with a learning rate $0 < \alpha < 1$ by

$$Q^{\pi,u}(s_t, a_t) = Q^\pi(s_t, a_t) + \alpha(y_t - Q^\pi(s_t, a_t)) \tag{22}$$

where y_t is the Temporal Difference (TD) target for $Q^\pi(s_t, a_t)$ and can be determined by

$$y_t^{q-learning} = R_t^{(n)} + \gamma^n \max_{a_{t+n}} Q^\pi(s_{t+n}, a_{t+n})$$

$$y_t^{sarsa} = R_t^{(n)} + \gamma^n Q^\pi(s_{t+n}, a_{t+n}) \tag{23}$$

$$R_t^{(n)} = \sum_{i=0}^{n-1} \gamma^i r_{t+i}$$

The learning rate α controls the speed at which $Q^\pi(s_t, a_t)$ updates, a lager α allows a fast update but may oscillate over training epochs while a smaller α tends to reserve the old $Q^\pi(s_t, a_t)$ and thus may take longer time to train.

Value-based RL does not work well for continuous control problems with infinite-dimensional action space or high-dimensional problems because it is difficult to explore all the states in a large and continuous space and store them in a table. In such a case, policy-based RL can provide better solutions than value-based RL. By treating the policy π_θ as a probability distribution over state-action pairs parameterized by θ, policy parameters θ are updated to maximize the objective function $J(\theta)$, which can be

$$J(\theta) = E_{\pi_\theta}[Q^{\pi_\theta}(s,a)|\theta] = \iint_{s \in S, a \in A} \pi_\theta(a|s)Q^{\pi_\theta}(s,a)dsda \tag{24}$$

The optimum policy parameters θ are selected to maximize $J(\theta)$, with

$$\theta_{opt} = \underset{\theta}{argmax}J(\theta) \tag{25}$$

Policy-based RL tries to select the optimum actions by using the gradient of the objective function with respect to θ, which can be written as

$$\nabla_\theta J(\theta) = \iint_{s \in S, a \in A} Q^{\pi_\theta}(s,a)\nabla_\theta \pi_\theta(a|s)dsda$$

$$= \iint_{s \in S, a \in A} \pi_\theta(a|s)Q^{\pi_\theta}(s,a)\nabla_\theta \log \pi_\theta(a|s)dsda \tag{26}$$

$$= E_{\pi_\theta}[Q^{\pi_\theta}(s,a)\nabla_\theta \log \pi_\theta(a|s)]$$

where $Q^{\pi_\theta}(s,a)$ cannot be determined directly and thus Monto-Carlo method is used to sample $Q^{\pi_\theta}(s,a)$ from M trajectories and take the empirical average

$$\nabla_\theta J(\theta) \approx \frac{1}{M}\sum_{m=1}^{M} R_{t_m}\nabla_\theta \log \pi_\theta(a|s) = \nabla_\theta \left[\frac{1}{M}\sum_{m=1}^{M} R_{t_m} \log \pi_\theta(a|s)\right] \tag{27}$$

where R_{t_m} is given by Eq. (19). No analytical solution can be provided for Eq. (27), thus, the optimum solution of Eq. (25) can be obtained by stochastic gradient descent algorithms with

$$\theta_{t+1} = \theta_t + \alpha R_t \nabla_\theta \log \pi_\theta(a|s) \tag{28}$$

where R_t is an estimator of the mean of the reward function. Eq. (28) shows that policy gradient can learn a stochastic policy by the update of the parameters θ at each iteration. Thus, policy-based RL does not need to implement an exploration and exploitation trade off. A stochastic policy allows the agent to explore the state space without always taking the same action. However, policy-based RL typically converges to a local optimum rather than a global optimum.

Actor-critic RL combines the characteristics of policy-based methods and value-based methods, in which an actor is used to control the agent's behaviors based on policy, critic evaluates the taken action based on value function. From Eq. (27), the objective function can be rewritten as

$$J(\theta) \approx \frac{1}{M}\sum_{m=1}^{M} R_{t_m} \log \pi_\theta(a_{t_m}|s_{t_m}) \tag{29}$$

The loss function for policy and value updating can be respectively defined as

$$L(\theta) = -\frac{1}{M}\sum_{m=1}^{M} R_{t_m} \log \pi_\theta(a_{t_m}|s_{t_m})$$

$$L(w) = \frac{1}{M}\sum_{m=1}^{M} (R_{t_m} - V_w(s_{t_m}))^2 \tag{30}$$

where $V_w(s_{t_m})$ is the approximation function for $Q^{\pi_\theta}(s_{t_m}, a_{t_m})$ and w is its parameter. Actor-critic RL aims to iteratively find the optimum θ and w to minimize the objective functions in (30).

Recall that R_{t_m} is obtained by taking T samples from the stored minibatch with $R_{t_m} = \sum_{i=0}^{T-1}\gamma^i r_{t_m+i}$, thus it may have a relatively large bias and variance. Advantage Actor-Critic (A2C) aims to solve the problem by learning the Advantage values A_{t_m} instead of R_{t_m}, which is defined by

$$A_{t_m} = R_{t_m} + \gamma^T V_{w^-}(s_{t_m+T}) - V_{w^-}(s_{t_m}) \tag{31}$$

where w^- is the parameter of the approximation function for the last iteration. Then, the loss function for policy updating can be rewritten by

$$L(\theta) = -\frac{1}{M}\sum_{m=1}^{M} A_{t_m} \log \pi_\theta(a_{t_m}|s_{t_m}) \tag{32}$$

3.2 Multi-agent deep reinforcement learning based traffic signal control

A real traffic network consists of multiple signalized intersections, each of which can be considered as an agent. The states for the i–th agent (intersection) such as the total number of approaching vehicles, position and speed of each vehicle, the vehicle flow and density of the links, etc. are not only determined by the i-th action (adjustment of green-time proportion) but also influenced by other agents' actions. Hence, traffic signal control can be modeled as a cooperative and competitive game, where the learning process requires considering other agents' actions to reach a globally optimum solution. When multiple agents are presented, the standard MDP is no longer suitable for describing the environment because actions from other agents can influence the state dynamics. In such a case, a Markov game can be defined by the tuple $(N, \{S_i\}_{i\in N}, \{A_i\}_{i\in N}, T, \{\pi_{\theta_i}\}_{i\in N}, \{r_i\}_{i\in N}, \gamma)$, where

- N is the number of agents with $N > 1$;

- A set of states space of the i-th agent S_i, which contains all possible states the i-th agent can be in and $S = S_1 \times S_2 \times \cdots \times S_N$ is called the joint state space;

- A set of action space of the i-th agent A_i and $A = A_1 \times A_2 \times \cdots \times A_N$ is called the joint action space;

- Transition probability $T(s_{t+1,i}|s_{t,i}, a_{t,1}, a_{t,2}, \cdots, a_{t,N})$ of the i-th agent: the probability transiting from $s_{t,i}$ to $s_{t+1,i}$ by a joint action $a = [a_{t,1} \quad a_{t,2} \quad \cdots \quad a_{t,N}] \in A$;

- Control policy of the i-th agent π_{θ_i}: the probability selecting an action by a given state and $\pi_\theta = [\pi_{\theta_1} \quad \pi_{\theta_2} \quad \cdots \quad \pi_{\theta_N}]$ is global control policy;

- The instantaneous reward function of the i-th agent r_i;

- $\gamma \in [0, 1]$ is the discount factor.

The centralized multi-agent RL considers the multi-agent systems as a single-agent system with joint state space S and action space A and thus it can achieve the global optimum. However, it is infeasible for large-scale traffic control systems because of the extremely high dimension of joint state-action space [32]. Decentralized Multi-Agent deep RL (MARL) can overcome the scalability issue by distributing the global control to each local agent, which is controlled based on the local observed states and communicated states and actions from other agents.

Suppose we have a multi-intersection traffic network, which can be modeled as $G(V, E)$. The size of the graph is the number of intersections denoted by N, and $e_{ij} = 1$ if vertices v_i and v_j are neighbors. The neighborhood of v_i can be manually selected within a certain coverage limit, denoted by Ω_i and thus the local region is defined by $L_i = \Omega_i \cup v_i$. Thanks to the advancement in communication technologies for ITS, it is possible to share the instantaneous rewards $\{r_i\}_{i\in N}$, states $\{s_{t,i}\}_{i\in N}$, actions $\{a_{t,i}\}_{i\in N}$ and policies $\{\pi_{\theta_i}\}_{i\in N}$ among agents. The objective of MARL is to maximize the total reward function $Q^{\pi_\theta} \approx \sum_{i=1}^{N} Q_i^{\pi_\theta}$, where $Q_i^{\pi_\theta}$ is the local reward of the i-th agent by given global control policy and can be expressed by

$$Q_i\left(s_{t,i}, a_{t,i}, \pi_{\theta_i,t}, s_{t,j=1,\cdots,N,j\neq i}, \pi_{\theta_j,t,j=1,\cdots,N,j\neq i}\right) = E\left[\sum_{t=0}^{T-1} \gamma^t r_{t,i}\Big| s_t, \pi_{\theta,t}\right]$$

$$= E\left[\sum_{t=0}^{T-1} \gamma^t r_{t,i}\Big| \underbrace{s_{t,i}, a_{t,i}, \pi_{\theta_i,t}}_{\text{localstateandpolicy}}, \underbrace{s_{t,j=1,\cdots,N,j\neq i}, \pi_{\theta_j,t,j=1,\cdots,N,j\neq i}}_{\text{communicatedstatesandpolicies}}\right]$$

$$= E\left[r_{t,i} + \gamma Q_i\left(s_{t+1,i}, \pi_{\theta_i}(s_{t+1,i}), s_{t+1,j=1,\cdots,N,j\neq i}, \pi_{\theta_j,t+1,j=1,\cdots,N,j\neq i}\right)\right] \quad (33)$$

where the global states and policies can be communicated from all other agents in the system as well as the neighborhood L_i. In multi-agent traffic signal control problems, the summation of each local reward can only be used to approximate Q^{π_θ} because of competitiveness among different intersections, i.e., a decrease of the average vehicle waiting time at the i-th intersection may cause waiting time increase at the neighboring intersections. Hence, the objective of decentralized MARL can be approximated to be the maximization of each local reward function with consideration of states and policies from the neighborhood, which can be expressed by

$$\theta_i^* = \underset{\theta_i}{argmax}\left[Q_i\left(s_{t,i}, a_{t,i}, \pi_{\theta_i,t}, s_{t,j=1,\cdots,N,j\neq i}, \pi_{\theta_j,t,j=1,\cdots,N,j\neq i}\right)\right] \quad (34)$$

We assume Eq. (33) has continuous state-action space and thus multi-agent A2C can be applied to search the optimum policy parameter. From Eq. (31), the Advantage value for the i-th intersection can be expressed by

$$A_{t_m,i} = R_{t_m,i} + \gamma^T V_{w_i^-}\left(s_{t_m+T}\big|\pi_{\theta_i^-}, \pi_{\theta_j^-,j=1,\cdots,N,j\neq i}\right) - V_{w_i^-}\left(s_{t_m}\big|\pi_{\theta_i^-}, \pi_{\theta_j^-,j=1,\cdots,N,j\neq i}\right)$$

$$= R_{t_m,i} + \gamma^T V_{w_i^-}(s_{t_m+T,L_i}) - V_{w_i^-}(s_{t_m,L_i})$$

$$(35)$$

where $R_{t_m,i}$ is the weighted sum of T samples from the minibatch, $V_{w_i}\left(s_{t_m}|\pi_{\theta_i^-}, \pi_{\theta_j^-, j=1,\cdots,N, j\neq i}\right)$ is the approximation of the value function for the input sample s_{t_m} with the function parameter of last time iteration, $T+1$ samples are obtained from the same stationary policy $\left[\pi_{\theta_i^-}, \pi_{\theta_j^-, j=1,\cdots,N, j\neq i}\right]$ which respectively represents the control policies for the i-th intersection and neighboring intersections at last time iteration, $s_{t_m,L_i} := \{s_{t_m,j}\}_{j\in L_i}$ represents the states for all neighbors of the i-th intersection at t_m. We assume that agents are synchronized, that is, the policy and value updating for all agents happen simultaneously at end of each episode, and the delay for information exchange among agents is ignored. Therefore, within each episode, the dynamic system can be considered to be stationary although the trajectory for each agent is influenced by multiple policies $\left[\pi_{\theta_i^-}, \pi_{\theta_j^-, j=1,\cdots,N, j\neq i}\right]$. Then, the loss function for policy and value updating can be obtained as

$$L(\theta_i) = -\frac{1}{M}\sum_{m=1}^{M} A_{t_m,i} \log \pi_{\theta_i}(a_{t_m,i}|s_{t_m,L_i}) L(w_i) = \frac{1}{M}\sum_{m=1}^{M}(R_{t_m,i} - V_{w_i}(s_{t_m,L_i}))^2 \quad (36)$$

If each agent follows Eqs. (35) and (36) in a decentralized manner, a local optimum policy $\pi_{\theta_i^*}$ can be achieved if other agents can achieve the optimum policy $\pi_{\theta_j^*, j=1,\cdots,N, j\neq i}$ within the same episode. However, if $\theta_j^*, j=1,\cdots,N, j\neq i$ cannot be achieved or $\theta_j, j=1,\cdots,N, j\neq i$ are updated within the same episode, the policy gradient may be inconsistent across minibatch and thus the convergence to a local optimum cannot be guaranteed, since $A_{t_m,i}$ is conditioned on the changing $\left[\pi_{\theta_i}, \pi_{\theta_j, j=1,\cdots,N, j\neq i}\right]$.

In practice, the information exchange among multiple intersections may not be synchronized and communication delay should be considered, which causes policy changing within the same episode and thus leads to non-stationarity. There is some research that try to stabilize convergence and relieve non-stationarity. Tesauro proposes a "Hyper-Q" learning, in which values of mixed strategies rather than base actions are learned and other agents' strategies are estimated from observed actions via Bayesian inference [55]. Foerster et al. include low-dimensional fingerprints, such as ε of ε-greedy exploration and the number of updates.

To relieve non-stationarity, the key is to keep policies from neighboring agents fixed within one episode. We can apply a DNN network to approximate the local policy $\pi_{\theta_i,t_m}(\cdot|s_{t_m,L_i})$ when size of L_i is relatively large. If we consider the sampled latest policies from neighbors to be additional input of the DNN network, besides the current state from L_i, the local policy can be rewritten by

$$\pi_{\theta_i,t_m} = \pi_{\theta_i}\left(\cdot|s_{t_m,L_i}, \pi_{\theta_j^-,t_m-1,j=1,\cdots,|L_i|,j\neq i}\right) \quad (37)$$
$$i = 1, 2, \cdots, N$$

Then, the loss function for policy updating can be rewritten by

$$L(\theta_i) = -\frac{1}{M}\sum_{m=1}^{M} A_{t_m,i} \log \pi_{\theta_i}\left(a_{t_m,i}|s_{t_m,L_i}, \pi_{\theta_j^-,t_m-1,j=1,\cdots,|L_i|,j\neq i}\right) \quad (38)$$

Even if the policies from the neighbors are fixed and are considered to be additional input, it is still difficult to approximate $A_{t_m,i}$ given by Eq. (35) and thus convergence to a local optimum cannot be guaranteed. Recall that the total reward function can be approximated as the summation of local reward functions. Thus, a decomposable global reward with a spatial discount factor can be proposed to solve the problem.

$$\tilde{r}_{t_m,i} = \sum_{j \in L_i | d(i,j) = d} \alpha^d r_{t_m,j} \tag{39}$$

where α is the spatial discount factor with $0 \le \alpha \le 1$, d is the distance between the i-th agent and j-th agent. The spatial discount factor scales down the reward in spatial order to emphasize the role played by the policy of the local agent. Compared to sharing the same weights across agents, the spatial discounted factor is more flexible for the trade-off between greedy control ($\alpha = 0$) and cooperative control ($\alpha = 1$), and is more relevant for estimating the advantage of local policy. By applying the spatial discount factor to neighboring states, we have

$$\tilde{s}_{t_m,L_i} = s_{t_m,i} \cup \alpha \left[s_{t_m,j} \right]_{j \in L_i} \tag{40}$$

Then, the cumulative discounted reward can be obtained by

$$\hat{R}_{t_m,i} = \sum_{t_m=0}^{T-1} \gamma^t \tilde{r}_{t_m,i} \tag{41}$$

and the local return and Advantage value $A_{t_m,i}$ for the i-th agent can be expressed by

$$\tilde{R}_{t_m,i} = \hat{R}_{t_m,i} + \gamma^T V_{w_i^-} \left(\tilde{s}_{T,L_i}, \pi_{\theta_j^-,T-1j=1,\cdots,|L_i|,j \ne i} \right)$$
$$A_{t_m,i} = \tilde{R}_{t_m,i} - V_{w_i^-} \left(\tilde{s}_{t_m,L_i}, \pi_{\theta_j^-,t_m-1j=1,\cdots,|L_i|,j \ne i} \right) \tag{42}$$

and Eq. (38) can be rewritten as

$$L(\theta_i) = -\frac{1}{M} \sum_{m=1}^{M} \left[\tilde{R}_{t_m,i} - V_{w_i^-} \left(\tilde{s}_{t_m,L_i}, \pi_{\theta_j^-,t_m-1j=1,\cdots,|L_i|,j \ne i} \right) \right]$$
$$\times \log \pi_{\theta_i} \left(a_{t_m,i} | s_{t_m,L_i}, \pi_{\theta_j^-,t_m-1j=1,\cdots,|L_i|,j \ne i} \right) \tag{43}$$

The loss function for value updating can be expressed as

$$L(w_i) = \frac{1}{M} \sum_{m=1}^{M} \left(\tilde{R}_{t_m,i} - V_{w_i} \left(\tilde{s}_{t_m,L_i}, \pi_{\theta_j^-,t_m-1j=1,\cdots,|L_i|,j \ne i} \right) \right)^2 \tag{44}$$

The decreolized MA2C can overcome the scalability issue and achieve local optimum (Pareto Optimality). How to achieve the global optimum using a decentralized approach when the global reward function is non-convex in the future research direction.

Compared to decentralized MA2C, Nash Q-learning does not consider cooperation among agents and thus it has lower computational complexity but can only achieve the Nash equilibrium. Nash Q-learning aims to find the optimal global control policy π_θ by iteratively updating agents' actions to maximize their Q functions based on assuming Nash equilibrium behavior, that is:

$$Q_{t+1,i}(s_{t,i}, a_{t,1}, \cdots, a_{t,N}) = (1 - \alpha)Q_{t,i}(s_{t,i}, a_{t,1}, \cdots, a_{t,N}) +$$
$$\alpha \left[r_{t,i} + \gamma Q_{t,i} \left(s_{t+1,i}, a_{t+1,1}^*, \cdots, a_{t+1,N}^* | \pi_{\theta,t}^* \right) \right] \tag{45}$$
$$a_{t,n} \leftarrow a_{t+1,n}^*, t \leftarrow t + 1$$

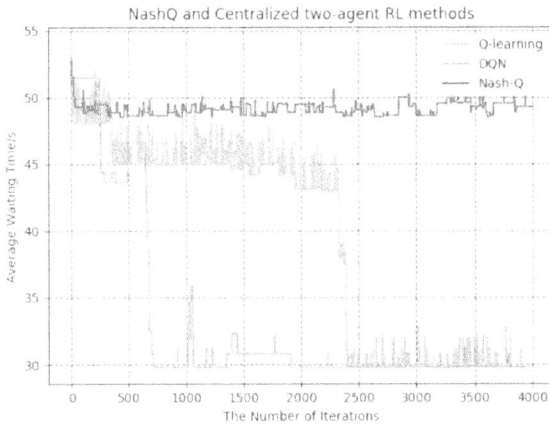

Figure 7.
Comparison of different multi-agent RL methods for traffic signal control.

where $Q_{t,i}(s_{t,i}, a_{t,1}, \cdots, a_{t,N})$ is the Nash Q function at the t-th iteration for the i-th agent, $\pi_{\theta,t}^*$ is the joint Nash equilibrium strategy at the t-th iteration. Under the joint Nash equilibrium strategy, the following relation should be fulfilled:

$$Q_{t,i}\left(s_{t+1,i}, a_{t+1,1}^*, \cdots, a_{t+1,N}^* | \pi_{\theta,t+1}^*\right) \geq$$
$$Q_{t,i}\left(s_{t+1,i}, a_{t+1,1}^*, \cdots, a_{t+1,N}^* | \pi_{\theta_1,t+1}^*, \cdots, \pi_{\theta_{i-1},t+1}^*, \pi_{\theta_i,t+1}, \pi_{\theta_{i+1},t+1}^*, \cdots, \pi_{\theta_N,t+1}^*\right) \tag{46}$$
$$\text{for } i = 1, 2, \cdots, N$$

Eqs. (45) and (46) show that at each iteration t, agent i observes its current state $s_{t,i}$ and takes action to maximize its Q function based on $s_{t,i}$ and other agents' actions. The update of i-th action will cause the update of actions of agents $-i$, which represents all agents excluding the agent i. The t-th joint Nash equilibrium strategy will not be obtained until the convergence of the joint Nash equilibrium action is reached.

In traffic signal control application, the state space S can be the number of vehicles, positions, speeds and lane order of vehicles, phase duration, etc., for the specific intersections, the action space A can be phase switch, phase duration or phase percentage, etc., the reward r can be queue length, average waiting time, cumulative time delay, network capacity etc. The objective of the multi-agent RL strategy is to minimize the accumulated average waiting for time or queue length etc.

By conducting a simulation on SUMO for a two-intersection case, we can observe in **Figure 7** that the centralized DQN outperform the centralized Q-learning in terms of reward value (Average Waiting Time/s) and convergence rate (the Number of Iterations). When the number of agents is small (two, in this case), by using the centralized methods, the average waiting time can converge to the local optimum, which is more optimal than the Nash equilibrium delivered by Nash Q learning. However, the convergence rate of Nash Q learning is higher than that of centralized methods.

4. Summary

In this chapter, we introduced deep learning-based traffic state prediction tech-nique, which can provide accurate future information for traffic control and decision making. The traffic state data depicts a strong correlation in the spatial and temporal domain, which can be utilized by applying CNN and LSTM techniques to improve the prediction accuracy. CNN technique is used to capture high-level spatial features while LSTM can provide excellent performance in dealing with time-sequential data by extracting high-level temporal features. We firstly reviewed the fundamentals of deep learning and presented the architecture of CNN and LSTM. Then, we introduced how to combine these two models to form concatenated hybrid models and parallelized hybrid models. Finally, we proposed bidirectional LSTM models to enhance prediction performance by learning additional high-level temporal features from the historical data in previous days.

Furthermore, we introduced the decentralized multi-agent advantage Actor-Critic technique and Nash Q learning for traffic signal control applications. We firstly briefly review the fundamental principles of RL. Then, we focus on multi-agent DRL-based traffic signal control techniques such as decentralized multi-agent advantage actor-critic, which can converge to the local optimum and overcome the scalability issue by considering the non-stationarity of MDP transition caused by policy update of the neighborhood.

The main contribution of this chapter can be summarized as followings:

- We reviewed the state-of-the-art technique in traffic state prediction and traffic control strategies, and provide readers with a clear framework for understanding how to apply deep learning models to traffic state prediction and how to deal with multi-agent traffic control by using RL strategies.

- We proposed the hybrid prediction models, which can utilize CNN and LSTM to capture the spatio-temporal feature of traffic data.

- We proposed a multi-agent deep RL (MARL) strategy, which conducts in a decentralized manner and considers the cooperation among agents and thus can overcome the scalability issue and achieve local optimum.

- We compared the centralized RL Q-learning, DQN to the Nash Q-learning strategy in terms of the reward value and convergence rate.

Author details

Shangbo Wang
Xi'an Jiaotong Liverpool University, Suzhou, China

*Address all correspondence to: shangbo.wang@xjtlu.edu.cn

IntechOpen

References

[1] Gora P, Rüb I. Traffic models for self-driving connected cars. Transportation Research Procedia. 2016;**14**:2207-2216

[2] Calvert S, Schakel WJ, Lint JWC. Will automated vehicles negatively impact traffic flow? Journal of Advanced Transportation. 2017;**2017**:8

[3] Yang Cheng YZ. Connected Automated Vehicle Highway (CAVH): A Vision and Development Report for Large Scale Automated Driving System (ADS) Deployment

[4] Liu Q, Li X, Yuan S and Li Z, Decision-Making Technology for Autonomous Vehicles Learning-Based Methods, Applications and Future Outlook, 2021

[5] Miglani A, Kumar N. Deep learning models for traffic flow prediction in autonomous vehicles: A review, solutions, and challenges. Vehicular Communications. 2019;**20**:100184

[6] Duan P, Mao G, Yue W, Wang S. A unified STARIMA based model for short-term traffic flow prediction. 21st International Conference on Intelligent Transportation Systems (ITSC). Maui, HI. 2018. pp. 1652-1657

[7] Napiah M, Kamaruddin I. ARIMA models for bus travel time prediction. Journal of the Institution of Engineers Malaysia. 2010;**71**

[8] Kumar SV, Vanajakshi L. Short-term traffic flow prediction using seasonal ARIMA model with limited input data. European Transport Research Review. 2015;7:21

[9] Williams BM, Durvasula PK, Brown DE. Urban freeway traffic flow prediction: Application of seasonal autoregressive integrated moving average and exponential smoothing models. Transportation Research Record. 1998;**1644**:132-141

[10] Ahn J, Ko E, Kim EY. Highway traffic flow prediction using support vector regression and Bayesian classifier. 2016 International Conference on Big Data and Smart Computing (BigComp). 2016. pp. 239-244. DOI: 10.1109/BIGCOMP.2016.7425919

[11] Kumar SV. Traffic flow prediction using Kalman filtering technique. Procedia Engineering. 2017;**187**:582-587

[12] Ranjan N, Bhandari S, Zhao HP, Kim H, Khan P. City-wide traffic congestion prediction based on CNN, LSTM and transpose CNN. IEEE Access. 2020;**8**:81606-81620

[13] Ma X, Dai Z, He Z, Ma J, Wang Y, Wang Y. Learning traffic as images: A Deep convolutional neural network for large-scale transportation network speed prediction. Sensors. 2017;**17**:818

[14] Di YANG, Songjiang LI, Zhou PENG, Peng WANG, Junhui WANG, Huamin YANG. MF-CNN: Traffic flow prediction using convolutional neural network and multi-features fusion. IEICE Transactions on Information and Systems. 2019;**E102.D**:1526-1536

[15] Bao X, Jiang D, Yang X, Wang H. An improved deep belief network for traffic prediction considering weather factors. Alexandria Engineering Journal. 2021;**60**:413-420

[16] Huang W, Song G, Hong H, Xie K. Deep architecture for traffic flow prediction: Deep belief networks with multitask learning. IEEE Transactions on Intelligent Transportation Systems. 2014;**15**:2191-2201

[17] Lu S, Zhang Q, Chen G, Seng D. A combined method for short-term traffic flow prediction based on recurrent

neural network. Alexandria Engineering Journal. 2021;**60**:87-94

[18] Sadeghi-Niaraki A, Mirshafiei P, Shakeri M, Choi S-M. Short-term traffic flow prediction using the modified elman recurrent neural network optimized through a genetic algorithm. IEEE Access. 2020;**8**:217526-217540

[19] Tian Y, Pan L. Predicting short-term traffic flow by long short-term memory recurrent neural network. 2015 IEEE International Conference on Smart City/SocialCom/SustainCom (SmartCity). 2015. pp. 153-158. DOI: 10.1109/Smart City.2015.63

[20] Wang W, Bai Y, Yu C, Gu Y, Feng P, Wang X, et al. A network traffic flow prediction with deep learning approach for large-scale metropolitan area network. 2018 IEEE/IFIP Network Operations and Management Symposium. 2018. pp. 1-9

[21] Wang S, Li C, Yue W, Mao G. Network capacity maximization using route choice and signal control with multiple OD Pairs. In: IEEE Transactions on Intelligent Transportation Systems. Vol. 21. 2020. pp. 1595-1611

[22] Keyvan-Ekbatani M, Kouvelas A, Papamichail I, Papageorgiou M. Exploiting the fundamental diagram of urban networks for feedback-based gating. Transportation Research Part B: Methodological. 2012;**46**:1393-1403

[23] Elouni M, Rakha HA. Weather-tuned network perimeter control - A network fundamental diagram feedback controller approach. 2018 International Conference on Vehicle Technology and Intelligent Transport Systems. 2018. pp. 82-90

[24] Haddad J, Shraiber A. Robust perimeter control design for an urban region. Transportation Research Part B: Methodological. 2014;**68**:315-332

[25] Sirmatel II, Geroliminis N. Economic model predictive control of large-scale urban road networks via perimeter control and regional route guidance. IEEE Transactions on Intelligent Transportation Systems. 2018;**19**:1112-1121

[26] Kouvelas A, Saeedmanesh M, Geroliminis N. A linear formulation for model predictive perimeter traffic control in cities**This research has been supported by the ERC (European Research Council) Starting Grant "METAFERW: Modelling and controlling traffic congestion and propagation in large-scale urban multi-modal networks" (Grant #338205). IFAC-PapersOnLine. 2017;**50**:8543-8548

[27] Aboudolas K, Geroliminis N. Perimeter and boundary flow control in multi-reservoir heterogeneous networks. Transportation Research Part B: Methodological. 2013;**55**:265-281

[28] Mohebifard R, Hajbabaie A. Optimal network-level traffic signal control: A benders decomposition-based solution algorithm. Transportation Research Part B: Methodological. 2019;**121**:252-274

[29] Wu W, Wang Z-J, Chen X-M, Wang P, Li M-X, Ou Y-J-X, et al. A decision-making model for autonomous vehicles at urban intersections based on conflict resolution. Journal of Advanced Transportation. 2021;**2021**:8894563

[30] Lu S, Liu X, Dai S. Incremental multistep Q-learning for adaptive traffic signal control based on delay minimization strategy. 2008 7th World Congress on Intelligent Control and Automation. 2008. pp. 2854-2858. DOI: 10.1109/WCICA.2008.4593378

[31] Shoufeng L, Ximin L, Shiqiang D. Q-learning for adaptive traffic signal control based on delay minimization strategy. 2008 IEEE International

Conference on Networking, Sensing and Control. 2008. pp. 687-691

[32] T. Chu, J. Wang, L. Codecà and Z. Li, "Multi-agent deep reinforcement learning for large-scale traffic signal control," IEEE Transactions on Intelligent Transportation Systems, PP. 2019

[33] Wang T, Cao J, Hussain A. Adaptive traffic signal control for large-scale scenario with cooperative group-based multi-agent reinforcement learning. Transportation Research Part C: Emerging Technologies. 2021;**125**:103046

[34] Wang X, Ke L, Qiao Z, Chai X. Large-scale traffic signal control using a novel multi-agent reinforcement learning. IEEE Transactions on Cybernetics. 2021;**51**(1):174-187

[35] Li Z, Xu C, Zhang G, A Deep Reinforcement Learning Approach for Traffic Signal Control Optimization, 2021

[36] Guo J, Harmati I. Evaluating semi-cooperative Nash/Stackelberg Q-learning for traffic routes plan in a single intersection. Control Engineering Practice. 2020;**102**:104525

[37] Shickel B, Tighe PJ, Bihorac A, Rashidi P. Deep EHR: A survey of recent advances on deep learning techniques for electronic health record (EHR) analysis. IEEE Journal of Biomedical and Health Informatics. 2018;**22**(5):1589-1604. DOI: 10.1109/JBHI.2017.2767063

[38] Lin C-T, Lee CSG. Neural Fuzzy Systems: A Neuro-Fuzzy Synergism to Intelligent Systems. USA: Prentice-Hall, Inc.; 1996

[39] Chon T-S, Park Y-S, Kim J-M, Lee B-Y, Chung Y-J, Kim Y. Use of an artificial neural network to predict population dynamics of the forest–pest pine needle gall midge (Diptera:

Cecidomyiida). Environmental Entomology. 2000;**29**:1208-1215

[40] Fouladgar M, Parchami M, Elmasri R, Ghaderi A. Scalable deep traffic flow neural networks for urban traffic congestion prediction. 2017 International Joint Conference on Neural Networks (IJCNN). 2017. pp. 2251-2258

[41] Yu H, Wu Z, Wang S, Wang Y, Ma X. Spatiotemporal recurrent convolutional networks for traffic prediction in transportation networks. 2018 Proceedings of the 2nd International Conference on Computer and Data Analysis (ICCDA). 2018. pp. 28-35

[42] Fang W, Love PED, Luo H, Ding L. Computer vision for behaviour-based safety in construction: A review and future directions. Advanced Engineering Informatics. 2020;**43**:100980

[43] Li H-C, Deng Z-Y, Chiang H-H. Lightweight and resource-constrained learning network for face recognition with performance optimization. Sensors. 2020;**20**

[44] Hassannayebi E, Ren C, Chai C, Yin C, Ji H, Cheng X, et al. Short-term traffic flow prediction: A method of combined deep learnings. Journal of Advanced Transportation. 2021;**2021**: 9928073

[45] Dinler ÖB, Aydin N. An optimal feature parameter set based on gated recurrent unit recurrent neural networks for speech segment detection. Applied Sciences. 2020;**10**

[46] Jagannatha A, Yu H. Structured prediction models for RNN based sequence labeling in clinical text. Proceedings of the Conference on Empirical Methods in Natural Language Processing. 2016;856-865

[47] Mohammadi M, Al-Fuqaha A, Sorour S, Guizani M. Deep learning for

IoT big data and streaming analytics: A survey. IEEE Communication Surveys and Tutorials. 2018;**20**:2923-2960

[48] Alzubaidi L, Zhang J, Humaidi AJ, Al-Dujaili A, Duan Y, Al-Shamma O, et al. Review of deep learning: Concepts, CNN architectures, challenges, applications, future directions. Journal of Big Data. 2021;**8**:53

[49] Hochreiter S, Schmidhuber J. Long short-term memory. Neural Computation. 1997;**9**:1735-1780

[50] Duan Z, Yang Y, Zhang K, Ni Y, Bajgain S. Improved deep hybrid networks for urban traffic flow prediction using trajectory data. IEEE Access. 2018;**6**:31820-31827

[51] Liu Y, Zheng H, Feng X and Chen Z. Short-Term Traffic Flow Prediction with Conv-LSTM. 2017

[52] Sainath TN, Vinyals O, Senior A, Sak H. Convolutional, Long Short-Term Memory, fully connected Deep Neural Networks. 2015 IEEE International Conference on Acoustics, Speech and Signal Processing (ICASSP). 2015. pp. 4580-4584

[53] Wu Y, Tan H. Short-term traffic flow forecasting with spatial-temporal correlation in a hybrid deep learning framework. ArXiv abs/1612.01022, 2016

[54] Haydari A, Yilmaz Y. Deep reinforcement learning for intelligent transportation systems: A survey. CoRR. 2020;abs/2005.00935

[55] Tesauro G. Extending Q-learning to general adaptive multi-agent systems. 2003 Proceedings of the 16th International Conference on Neural Information Processing Systems (NIPS). 2003. pp. 871-878

Chapter 5

Vehicle-To-Anything: The Trend of Internet of Vehicles in Future Smart Cities

Mingbo Niu, Xiaoqiong Huang and Hucheng Wang

Abstract

This chapter includes five parts—the concept of vehicle-to-anything (V2X), introduction of visible light communication (VLC), free-space optical communication (FSO), and terahertz (THz). The first part will present the concept of V2X. V2X is the basis and fundamental technology of future smart cars, autonomous driving, and smart transportation systems. Vehicle-to-network (V2N), vehicle-to-vehicle (V2V), vehicle-to-infrastructure (V2I), and vehicle-to-people (V2P) are included in V2X. V2X will lead to a high degree of interconnection of vehicles. The concept of VLC is presented in the second part. Intelligent reflecting surface (IRS) for nano-optics and FSO communication is introduced in the third part. At the same time, IRS keeps pace with the phase in communication links. Prospects of THz in glamorous cities are introduced in the fourth part. These new technologies will lead to trends in the future. A comparison of optical communication technology and applications in V2X is described in the fifth part.

Keywords: Vehicle-to-anything, visible light communication, free-space optical communication, smart city, terahertz technology

1. Part I: V2X Introduction

1.1 What is V2X?

V2X refers to the realization of a full range of network connections among V2V, V2P, V2I, and V2N with the help of new advances in information and communication to improve the level of intelligence and autonomous driving capabilities of the vehicle. **Figure 1** shows the components of V2X. On one hand, V2X will improve traffic efficiency; on the other hand, it will provide users with intelligent, comfortable, safe, energy-saving, and efficient integrated services. V2X will establish a new direction for the development of automotive technology by integrating global positioning systems, wireless communication, and remote sensing technologies [1]. At the same time, V2X will realize the compatibility of manual driving and automatic driving. In the automatic driving mode, it is possible for the automatic vehicle to randomly select the driving route with the best road conditions through the analysis of real-time traffic information. This driving mode can alleviate traffic jams. In addition, through the use of onboard sensors and cameras, vehicle can perceive the surrounding environment and make rapid adjustments to achieve "zero traffic

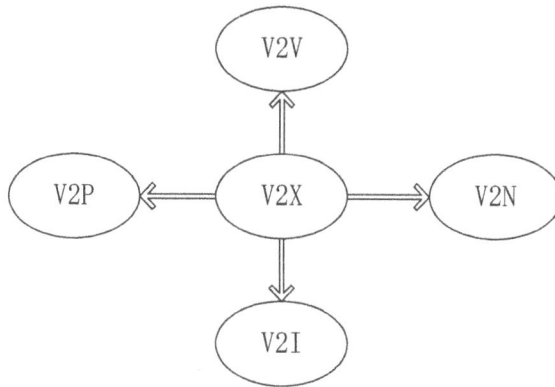

Figure 1.
The components of V2X.

accidents." For example, if a pedestrian suddenly appears, the car will automatically slow down to a safe speed or stop [2].

The earliest application of V2X was shown on a Cadillac by General Motors in 2006. Since then, other auto-product suppliers have begun to study this technology. However, the application of V2X was put on the agenda because of two traffic accidents that originated in the United States.V2X is the key technology of the future intelligent transportation system (ITS). V2X makes communication between vehicles and base stations easier. A series of messages, such as real-time road conditions, traffic signals, and pedestrian information, can be obtained. These messages can improve driving safety, reduce congestion, and improve traffic efficiency. The purpose of V2X is to reduce accidents, alleviate traffic congestion, reduce environmental pollution, and provide additional information services.

1.2 Motivation

If a vehicle can be illustrated as the driver's second pair of "eyes," it can theoretically reduce the occurrence of traffic accidents caused by driver's distraction or low visibility. V2X is a clever technology that turns the vehicle into the driver's eyes. V2X can see animals that suddenly run on the road before the driver notices. Generally, V2X uses neighbor cars to see traffic signal indicators and remind the driver while the driver can be hard to notice. Compared with cameras or Lidar commonly used in autonomous driving [3], V2X has the ability to break through visual blind spots and cross obstructions to obtain traffic information. At the same time, V2X shares real-time driving status with other vehicles or facilities and decides the driving state of the vehicle immediately through study and judgment algorithms information. In addition, V2X is free subject to extreme weather conditions, such as rain, fog, and strong light exposure. Therefore, V2X is being developed in transportation, especially in the field of autonomous driving.

1.3 V2X development status

In 2015, US launched the ITS five-year plan with the theme "change the way in which society moves forward." The main technical goals of planning are to "To realize application of connected vehicles" and "To accelerate autonomous driving."

Six categories of projects are defined in the plan—accelerated deployment, connected vehicles, autonomous driving, emerging capabilities, interoperability, and enterprise data. Connected vehicles, autonomous driving, and emerging capabilities are the three paths of technological development, while interoperability and enterprise data are the cornerstones of ITS development. To promote the further development of V2V and to reverse the subsequent legislative decisions of US, US Department of Transportation has led the "Safety Pilot Demonstration Deployment" project based on V2V and V2I. On the basis of the test and verification of the "Safety Pilot Demonstration Deployment" project, in 2014, the US National Highway Traffic Safety Administration announced the draft of the Vehicle-to-Vehicle Communication Advance Law, and launched the NPRM process in 2016 to enforce the light-duty vehicle V2V Communication, the main content includes:

1. Proposed mandatory V2V communication based on IEEE 802.11p

2. Specified the content of the BSM message

3. Specified V2V communication performance requirements

4. Specified privacy and security requirements

5. Designated equipment authorization system

The period from 2022 to 2025 is the deployment and development period of C-V2X industrialization. After 2025, the rapid development of the C-V2X industry will gradually achieve national coverage of C-V2X, and China will build a nationwide multi-level data platform, achieve cross-industry data interconnection, and provide diversified travel services.

1.4 V2N

V2N refers to the connection of vehicle devices with the network. Network exchanges data with the vehicle, stores and processes the acquired data, and provides various application services required by the vehicle. V2N communication is mainly used in vehicle navigation, remote vehicle monitoring, emergency rescue, and infotainment services.

1.5 V2V

V2V refers to communication between vehicles through onboard terminals. The vehicle-mounted terminal can obtain information, such as the speed, location, and driving conditions of surrounding vehicles in real time. Vehicles can also form an interactive platform to exchange information, such as pictures and videos in real time. V2V communication is mainly used to avoid or reduce traffic accidents, vehicle supervision, and management [4].

V2V enables sensors to communicate with neighboring vehicles, and it is more accurate and energy-efficient than any onboard surround sensing system. If we study further, we will come up with autonomous driving not a solution for the transition of automated vehicles from A to B, but a network protocol that optimizes traffic parameters and allows all commuters to reach their destination quickly and safely. Wireless upgrade is another basic autonomous driving function enabled by V2N. Since autonomous driving is a life-critical application, it must be kept up to the latest status.

1.6 V2I

V2I refers to the communication between vehicle-mounted equipment and roadside infrastructures, such as traffic lights, traffic cameras, and roadside units. The roadside infrastructure can also obtain information about vehicles in nearby areas and release various real-time information. V2I communication is mainly used in real-time information services, vehicle monitoring and management, and non-stop toll collection. V2I sensors collect information about traffic, traffic light status, radar equipment, cameras, and other road signals work as shared nodes to maximize infrastructure throughput. Even object-list lane markings or road barriers will one day become "smart" and become V2I communicators. For autonomous driving, information is critical because the vehicle may rely on stationary object data specifically for certain road events. Vehicles approaching the work area can notify and slow down. The parking lot can announce the availability of the previous passenger the moment they leave the scene [4].

In addition, vehicle can collect traffic data flow and help the driver choose the best route. Due to real-time traffic updates, V2I can reduce fuel consumption. Pre-filtering through V2I and autonomous driving can increase density, which will quadruple the current infrastructure capacity, keep road accidents at zero and increase traffic speed.

1.7 V2P

V2P means vulnerable traffic groups, including pedestrians and cyclists, use user equipment to communicate with vehicle-mounted devices. V2P communication is mainly used to avoid or reduce traffic accidents. By organically linking "people, vehicles, infrastructure, network" and other elements, V2X can not only support vehicles to obtain more information than bicycles perceive, promote the innovation and application of autonomous driving, but also help build a more intelligent environment. The transportation system promotes the development of new models and business of automobiles and transportation services. V2P is of great significance for improving traffic efficiency, saving resources, reducing pollution, reducing accident rates, and improving traffic management [5]. **Figure 2** shows the typical V2X scenario as follows.

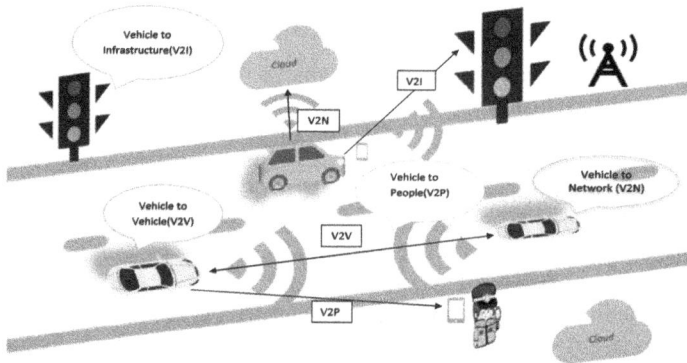

Figure 2.
Typical V2X scenario.

2. Part II: Visible light communication

2.1 Introduction of VLC

VLC refers to a type of communication that transmits data by modulating light waves in the visible spectrum (wavelength range from 380 nm to 750 nm). VLC is an emerging technology that realizes data communication by modulating the light intensity information emitted by light-emitting diode. Generally speaking, a system using visible light can be called VLC. VLC transmits data in a subtle way without affecting the normal lighting environment [6]. VLC is an optical communication technology, which uses optical transmission for information transmission. The spectrum is the physical threshold of light transmission, as shown in **Figure 3** is the spectrum of VLC [7].

Visible light usually uses LED as a communication medium. LED equipped electroluminescence and semiconductor to generate light, which is made by conducting materials. Due to high energy efficiency, durability, and low cost, LED sales have doubled. LED has been widely used in various devices, such as smartphones, vehicles, video screens, and signs. The universe of VLC has brought many benefits to the industry. LED bulbs have become the main medium for visible light communication.

2.2 Development of VLC in transportation system

So far, ITS relies on RF. However, the last decade has seen a major shift in lighting technology. With the major breakthrough of optical communication technology and the wide application of LED in indoor or outdoor lighting stimulation, VLC has become a feasible communication technology, making Vehicular VLC (V-VLC) possible in ITS [8].

A major advantage of VLC is the use of existing infrastructure to provide communication services. Data and energy can be transmitted simultaneously through LED. That is, energy transmitted does not increase the cost [9].

One advantage of visible light over radio frequencies is the size of the frequency spectrum. The allocation of frequencies in the radio frequency band of the electro-magnetic spectrum is greatly limited, regulated by each country, and coordinated through international telecommunications agencies. Light, however, is a different material. The spectrum of visible light is completely free, and it will lead to different commercials and academic possibilities [9].

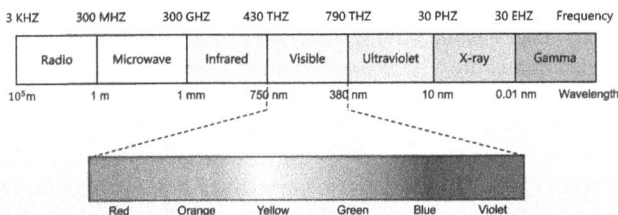

Figure 3.
The frequency band of visible light in the electro-magnetic spectrum [7].

Because of its propagation properties, light has a security advantage over radio waves. RF for vehicle-mounted networks has nondirectional propagation, relatively long communication distances, and it can penetrate objects. RF has been well studied over the past few decades and the technology is quite mature. But due to the potential security attacks, such as jamming eavesdropping, and man-in-the-middle attacks, it will raise concerns about their use in security-critical on-board networking applications. Light, on the other hand, does not follow this behavior. Light has high directivity, and its physical properties provide a more secure environment for communication systems [10].

Finally, one of the main advantages of light is the wave's high frequency, which allows for very high data rate communication. A large amount of bandwidth available in the visible spectrum allows for huge potential data rates. Currently, in terms of Wi-Fi, the highest data rates achieved in standard Wireless Gigabit are close to 1 Gbps [11]. Due to the high frequency of light waves, VLC searches have yielded impressive results, which speed up to 100 Gbps [12].

2.3 VLC modulation

VLC contains an irradiance modulation with direct detection (IM/DD) to communicate data faster than the persistence of human eyes by modulating LED intensity. Compared with traditional RF, VLC has superior speed and efficiency, security, and low cost. VLC fulfills its dual purpose of lighting and high-speed data communication. According to the characteristics of different modulation schemes, VLC modulation in visible light communication is divided into single carrier modulation multi-carrier modulation and Color Gamut-based Modulation.

2.3.1 Single carrier modulation

Single carrier modulation is the transmission of all data signals using a single signal carrier. Single carrier avoids the problem that the ratio of maximum instantaneous electric power to average electric power of a multi-carrier system at the same time of each phase is very large. This technology is more mature and the system has higher stability. For the best point-to-multipoint communication system, the single carrier modulation can make the frequency and time synchronization design easier, and improve the stability of the system. A single carrier modulation system provides a point-to-multipoint wireless communication solution with high efficiency, high flexibility, and high stability.

Common single-carrier modulation schemes include on–off Keying (OOK) and Pulse Position Modulation (PPM). OOK is a simple amplitude shift keying modulation. Because its modulation is simple and easy to implement, it is widely used in low- and medium-speed data rate demand application scenarios. Although the latest research has realized data transmission at 1250 Mbits over a distance of 1 m [13], transmission cannot be promoted due to the limitation of transmission distance.

PPM has been developed as an alternative communication technology to improve the anti-interference capability of information transmission. PPM is a good way of modulation [14]. Considering the bit error rate performance, bandwidth requirements, optical power, and optical implementation complexity, PPM modulation is a viable candidate for VLC communication. Compared to OOK, pulse position modulation has low noise interference because the amplitude and width of the pulse are constant during modulation. In pulse position modulation, noise removal and separation are very easy. Due to the constant pulse amplitude and width, the power consumption is also very low compared with other modulation methods.

In the past few decades, PPM modulation technology has received extensive attention, and research has been extended to various forms, such as differential pulse position modulation (DPPM), digital pulse interval modulation, multi-pulse position modulation, overlapping pulse position modulation (OPPM), and pulse rate modulation. DPPM is a simple improvement of PPM modulation. As long as deleting all the "0" time slots behind the "1" time of the PPM modulation, we can get the corresponding DPPM signal. Compared to PPM, DPPM symbols do not have strict symbol synchronization requirements, and more importantly, they can provide higher power utilization and bandwidth utilization. However, the bit error rate in DPPM is higher than that in PPM [15]. The main disadvantage of this scheme is that the pulse width is very short, and the high order M-element PPM modulation VLC can improve the power and bandwidth efficiency [16]. PPM modulation index can improve the power of the system from 1 dB to 2.5 dB by reducing the average bit error rate (BER) [16]. OPPM signal modulation schemes offer key advantages over other existing PPM schemes, such as greater sensitivity and smaller bandwidth expansion [17]. A priority decoding OPPM error correction scheme is proposed, which can significantly improve the system's BER without affecting the system bandwidth [14].

But PPM still has some disadvantages, for example, synchronization between transmitter and receiver, which is not always possible; we need dedicated channels and like pulse-amplitude modulation, transmission requires high bandwidth and this modulation requires special equipment. In addition, single carrier modulation is commonly subject to inter-symbol interference (ISI) during high-speed data transmission, which means that new modulation techniques are required.

2.3.2 Multi-carrier modulation

With the increase in the VLC network data rate, MCM was developed to solve ISI during high-speed data transmission. MCM divides the transmitted data stream into several different components through different sub-channels. Under ideal propagation conditions, subchannels are usually orthogonal, and the number of substrates is chosen so that the bandwidth of each subchannel is lower than the coherent bandwidth of the channel, thus making it relatively immune to flat fading. Compared with incoherent modulation, MCM has lower energy efficiency and higher bandwidth efficiency. Common multi-carrier modulation includes subcarrier intensity modulation (SIM) and orthogonal frequency division multiplexing (OFDM), which have the advantages of high spectral efficiency and recovery ability against channel damage. However, SIM modulation is mostly used in the study of FSO, and here we discuss more OFDM [18].

OFDM modulation solves the multi-user problem by dividing the parallel data stream into different narrowband channels at different frequencies. However, most VLC systems use IM/DD, which requires that the electrical signal must be a real positive signal, so baseband OFDM cannot be directly applied. The improved schemes for OFDM, VLC include direct current (DC) bias optical OFDM (DCO-OFDM) and asymmetric limiting optical OFDM (ACO-OFDM). In the DCO-OFDM system, a DC offset is added to the normal OFDM symbol to reduce signal distortion and noise caused by negative limiting. In ACO-OFDM, only odd-indexed subcarriers are modulated, and the negative signal is clipped to zero during transmission. Compared with ACO-OFDM, DCO-OFDM has a lower power efficiency, but higher spectral efficiency. With the increase of the modulation order, BER performance of ACO-OFDM is about 4.5 dB better than that of DCO-OFDM, reaching 10^{-3} [19]. In the case of small bias, BER of ACO-OFDM with 16 quadrature amplitude modulation (QAM) format is lower than that of ADO-OFDM with 4 QAM format.

2.3.3 Color-shift keying based on color gamut

Color-shift keying (CSK) is a visible light communication intensity modulation scheme proposed in IEEE 802.15.7, which sends signals through the color intensity emitted by red, green, and blue (RGB) light-emitting diodes. CSK signal points can be represented by an intensive combination of RGB colors corresponding to the transmitted data. The flicker of the light source is reduced by keeping the total emission intensity constant. Due to its unique advantages in preventing scintillation effect and light intensity fluctuation, the research of CSK in vehicle-mounted VLC has attracted more and more attention in recent years.

RGB LED consists of three LEDs in a package and produces white light through a combination of red, green, and blue outputs. Although more costly, RGB LED light can produce any perceived lighting color and can increase VLC data throughput by acting as a separate communication band. The perceived lighting color can be modified while achieving higher spectral efficiency.

3. Part III: Free-space optical communication

3.1 FSO introduction

FSO is known as an optical wireless system, which uses the atmosphere between the transmitter and receiver as the propagation medium, and FSO communication link is a line of sight (LOS). It can be used on various platforms, such as satellites, ships, airplanes, and other stationary or moving space and atmosphere. Due to its unregulated spectrum, inherent security, high data rate, and wider bandwidth, FSO is considered to be a supplement to radio frequency communication [20]. Although optical communication has advantages that traditional communication links cannot match, because its propagation medium cannot be controlled or adjusted, optical communication systems will be affected by some atmospheric phenomena. The main challenges of FSO are narrow beamwidth, transmission signal scattering, and scintillation. Due to the influence of atmospheric turbulence, the received signal intensity will fluctuate, that is, flicker [21]. FSO works according to the principle of sight. For continuous data transmission, LOS generated by the light beam should be straight. FSO is a combination of wireless technology and optical technology. The main factor that needs to be considered is the optimal bandwidth of the light beam used for communication and transmission of information signal data, such as audio and video. Free-space technology is an older technology used for lower data communications. Due to the limited bandwidth, RF is limited. Using lighting to transmit data, Li-Fi technology was proposed in optical communication [22]. The wireless optical communication technology was developed by the National Aeronautics and Space Administration and used for military purposes with high-speed communication links [23].

In recent years, FSO have received extensive attention in terms of ground-to-satellite transmission links and last-mile applications due to their high capacity and easy implementation. However, atmospheric turbulence can cause random fluctuations in the amplitude and phase of the received signal, which limits the application of FSO links. Multiple-input multiple-output (MIMO), adaptive optics, and fiber laser phased array (FLPA) are important ways to suppress atmospheric turbulence [24].

3.2 Performance of FSO

The performance analysis of FSO should be considered from external and internal parameters. The specifications and ratings of the components used, including operating frequency, divergence, power consumption, and transmission angle are all internal parameters. The ability of the lens and the error rate are all at the receiver end. External parameters include environmental factors, such as alignment, atmospheric attenuation, weather conditions, and scintillation.

FSO communication depends on weather conditions. If weather conditions are cloudy or visibility is lower, the formed communication link would not be sufficient for effective communication, whereas the performance of FSO relies on the weather conditions. In the FSO system, the transmitter would produce a narrow beam of light, and the narrow beam of light is straight. At the same time, the receiver should receive the narrow beam of light from the strong communication link on a straight line [25]. In the optical communication system, a straight beam with a diameter of 5–8 cm passes through and spreads to 1–5 m within 1 km.

3.3 FSO efficiency

FSO technology is changing rapidly day by day. This technology would increase and maximize signal bandwidth, at the same time, this technology transmitting data would be at high speeds. FSO technology is similar to fiber optical communication. The only conversion is signal path flow. That is to say, wireless communication between the transmitter and receiver, without cables, so it reduces costs and can be more efficient [26]. The efficiency of FSO mainly depends on the external aspects or the medium aspects between the transmitter and the receiver. Data transmission is lossless and high-speed if the transmission medium has strong visibility. The data transmission speed of the LED can reach 100 Mbps, and various experiments have been carried out to increase the data rate.

A fiber laser phased array transmitter into a FSO communication system and compared BER and optical transmit power of the two systems, which used single-aperture transmitter and FLPA transmitter. Experiments show that the power budget gap is about 8–10 dBm [24]. This shows that the FLPA transmitter provides a higher power budget. A new type of FSO switch capable of multicasting, the cost analysis of this switch shows that even if the cost of T-SE is 1.2 to 3.5 of micro-electro-mechanical system mirroring, its cost is lower than that of AD-based switches [27]. In Ref. [16], this paper uses avalanche photodiode (APD) and the positive intrinsic negative (PIN) receivers, respectively, and considers a single input multi output system with strong gas turbulence defined by M-ary PPM modulation and gamma-gamma distribution. Then a comprehensive comparative evaluation of the two situations is carried out. The experimental results show that the performance of the system can be improved by increasing the strength. In addition, we compared the main parameters of FSO and RF in order to distinguish the differences between them more intuitively, as shown in **Table 1**.

3.4 What is IRS?

As a new invention, IRS can be called smart wall, smart reflective light, passive smart mirror, smart reflective surface, and large smart surface. IRS is composed of a

Parameters	FSO	RF
Light source characteristics	Laser communication and non-laser communication	Radio frequency identification
Modulation rate	high	low
Anti-interference ability	low	high
Power	2.00E-03(J/Mb)	2.31E-02 (J/Mb)
Power loss	5–15 db/km	108 dB/km
Output power	5–500 mWatt	50 mWatt
Range	4 km	4 km
Data rate	10 Gbps	100 Mbps
Capacity	Not Allowed	Allowed
Advantage	Unlicensed band	NLOS
Security	high	low
Limitation	environment	spectrum
Spectrum range	0.8–1.2 THz	2–6 GHz
Wavelength	850 nm–1550 nm	22–2500 m

Table 1.
The performance comparison between FSO and RF.

large number of passive, low-cost components. It is a low-carbon and environmentally friendly smart component that can effectively control the phase, frequency, amplitude, and even polarization of the collision signal, IRS will build a real-time and reconfigurable propagation environment. The signal coverage of IRS is small, easy to deploy, and will not interfere with each other. By increasing the number of reflective elements, the quality of the received signal can be significantly improved. IRS does not require a power supply, complex algorithms, and hardware. IRS is easy to integrate into current wireless communication systems. These advantages make IRS a promising candidate for future wireless communication systems. IRS can greatly adjust the signal reflection to change the wireless channel to enhance communication performance. IRS is used to realize the intelligent and reconfigurable wireless channel propagation environment of the B5G/6G wireless communication system. Generally speaking, IRS is a plane composed of a large number of passive reflection units, and each passive reflection unit can independently produce a controllable amplitude and/or phase change of the incident signal. By densely deploying IRS units in the wireless network, the reflection of the IRS array is cleverly coordinated. The signal propagation between the transmitter and the receiver can be flexibly reconfigured to achieve the required realization and distribution, which provides a new means to fundamentally solve the problem of wireless channel fading damage and interference. It is possible to achieve a leap in wireless communication capacity and reliability.

3.4.1 Features of IRS

1. **Passive**

   IRS is composed of a large number of low-cost passive reflective components, which are only used to reflect signals and do not need to transmit signals. Therefore, IRS is almost passive and ideally does not require any dedicated energy.

2. **Programmable control**

IRS can control the scattering, reflection, and refraction characteristics of radio waves through the program, thereby overcoming the negative effects of natural wireless propagation. Therefore, IRS-assisted wireless communication can intelligently control the wave-front, such as phase, amplitude, frequency, and even polarization, which can hit the signal without complicated decoding, encoding, and radio frequency processing operations.

3. **Good compatibility**

IRS can be integrated into the existing communication network protocol only by changing the network, without changing the hardware facilities and software of their equipment. At the same time, the IRS has a full-band response, and it can ideally work at any operating frequency.

4. **Easy to deploy**

IRS is characterized by small size, lightweight, conformal geometry, and thinner than the wavelength, so it is easier to install and disassemble. Therefore, IRS can be easily deployed on exterior walls of buildings, billboards, ceilings of factories and indoor spaces, and people's clothes.

3.4.2 Application of IRS in V2X

In the future, IRS will be everywhere. IRS can be deployed on outdoor walls, drones, and transportation equipment of smart buildings in smart cities. Self-driving vehicles can use IRS as an intermediate medium to realize free wireless optical transmission, quickly, accurately, and accurately convey various information to the vehicle, and realize V2X. Centralizing vehicles into the Internet of Things makes the vehicle and the Internet of Things closely connected. The lightweight, convenient, and flexible deployment characteristics of IRS enable IRS to play a big role in V2X.

Another promising direction is IRS-assisted RF sensing and positioning. The large aperture size of the IRS and its ability to shape the propagation environment can significantly enhance RF sensing capabilities. The channel can be changed to provide favorable conditions for RF induction, and then it can be monitored with high precision. Encouraging results were reported in Ref. [28], and these results may have applications in energy-saving monitoring, assisted living, and remote health monitoring. However, the issue of optimizing the configuration of the IRS to enhance RF sensing remains to be studied. The effective combination of radio frequency technology and IRS can also be applied in the future V2X. Vehicles can use sensors to sense the signals of surrounding vehicles and traffic signs and provide evidence of effective traffic information for real-time V2X decision analysis. The combination of RF and IRS makes the monitoring data correct.

4. Part IV: Terahertz technology

4.1 Introduction to THz

In the past few years, wireless data traffic has seen unprecedented growth. On one hand, from 2016 to 2021, mobile data traffic is expected to increase seven times. On the other hand, video traffic is expected to triple in the same period [29]. In fact, by 2022, wireless and mobile device traffic is expected to account for 71% of total traffic. In fact, by 2030, wireless data rates will be sufficient to match wired broadband Competition [30]. The growth of use of wireless communication has led the researcher to explore appropriate radio spectrum ranges to satisfy the growing needs of individuals. For this reason, THz frequency band (0.1–10 THz) has begun to attract attention. Seamless data transmission, unlimited bandwidth, microsecond delay, and ultra-high-speed downloading of THz will completely lead the innovation of communication and change the way of communication and access information.

The term terahertz was first used in the field of microwave science in the 1970s to describe the spectral frequency of interferometers, the coverage of diode detectors, and water laser resonance [31, 32]. In 2000, terahertz was called a submillimeter wave, and the frequency range was between 100 GHz and 10 THz. However, the dividing line between sub-millimeter wave and far infrared was not clearly identified [33, 34]. The concept of ultra-wideband communication using THz for no line of sight signal components was first proposed as a powerful solution for extremely high data rates [35]. Since then, THz technology, especially communication technology, has captured the enthusiasm of the research community.

In fact, the rise of terahertz wireless communication started as early as 2000 when the 120 GHz wireless link produced by photonic technology started [36]. The 120 GHz signal is the first commercial terahertz communication system, and its allocated bandwidth is 18 GHz. Data rates of 10 gbps and 20 gbps are achieved through OOK or QPSK modulation, respectively [37, 38]. The terahertz frequency band guarantees a wide range of throughput, and theoretically can be extended to several terahertz to reach terabits per second (Tbps) [39]. This potential associated with terahertz technology has attracted a wider research community. In fact, the joint efforts of active research teams are producing new designs, materials, and manufacturing methods, providing unlimited opportunities for the development of terahertz. The potential benefits of the THz band are discussed [33]. THz can be applied to terahertz imaging and tomography [34]. THz wave is an electromagnetic wave between microwave and infrared, with a wavelength of 0.03–3 mm and a frequency of 0.1–10 THz. THz waves not only have the same straightness as light waves but also have similar penetrating and absorptive properties to radio waves.

4.2 Characteristics of THz radiation

1. Transient: The typical pulse width of a THz pulse is in the order of picoseconds.

2. Broadband: THz pulse source usually only contains several periods of electromagnetic oscillation, and the frequency band of a single pulse can cover the range from GHz to tens of THz.

3. Coherence: The coherence of THz comes from its generation mechanism.

4. Low energy: The energy of THz photons is only Millielectron volts. Compared with X-rays, it will not damage the detected substance due to ionization.

5. Penetration: THz radiation has a strong penetrating power for many nonpolar materials, such as dielectric materials and plastics, cartons, and other packaging materials. It can be used for quality inspection of packaged items or for a safety inspection.

Most polar molecules, such as water molecules and ammonia molecules, have strong absorption of THz radiation. The spectral characteristics of THz can be analyzed to study material composition or perform product quality control.

4.3 Problems with THz

Because most biological tissues are rich in water, water absorbs THz radiation very strongly, it greatly reduces the sensitivity of imaging of biological samples, and THz cannot make a clear image for samples with a lot of water, especially thick samples. This severely limits the application of THz imaging in biomedicine.

At present, the average energy of THz waves generated by most femtosecond lasers is only on the order of Nanowatts and can reach a signal-to-noise ratio of 100,000 or higher for single-point detection, but the signal-to-noise ratio of real-time two-dimensional imaging is very low. To obtain a high signal-to-noise ratio for imaging, a higher energy source is required.

4.4 Future research direction of THz

4.4.1 Terahertz ultra-massive MIMO

THz ultra-large MIMO frequency band can meet the needs of high data rates, but under the premise of providing a huge bandwidth, this band suffers a huge atmospheric loss. Therefore, high-gain directional antennas should be used for communication over a distance of more than a few meters. In the terahertz band, antennas are installed in the same space in a small and dense manner. Ultra Massive MIMO (UM-MIMO) channel was proposed [40, 41], the concept of UM-MIMO relies on the use of ultra-dense frequency-tunable plasma nano-antenna arrays, UM-MIMO was used for both transmitting and receiving, thereby increasing the communication distance and ultimately increasing the achievable data rate at terahertz frequencies [42]. In fact, when ensuring a two-dimensional or planar antenna array instead of a one-dimensional or linear array, the radiated signal can be adjusted in elevation and azimuth directions. This results in 3D or full-dimensional MIMO. The performance of UM-MIMO technology depends on two indicators, namely the prospect of plasmonic nano-antennas and the characteristics of the terahertz channel. Another important aspect is dynamic resource allocation, which can make full use of the UM-MIMO system and obtain maximum benefits through adaptive design schemes [43].

4.4.2 Terahertz virtual reality perception through cellular networks

Facing the technical barriers of 5G communication, THz is expected to have breakthroughs in reliability and low latency. Currently, video requires extremely high bandwidth. Therefore, the terahertz frequency band is sought as a technological supplement, and THz will provide high capacity and dense coverage to meet user needs. The terahertz cellular network will enable interactive, high dynamic range video with higher resolution and higher frame rate, which actually requires 10 times the bit rate required for 4 K video. Terahertz transmission will help solve any interference problems and provide additional data to support various instructions in video transmission. In addition, the terahertz band will become an enabler of 6-degree-of-freedom (6DoF) video, providing users with the ability to move inside and interact with the environment. The results of the literature [44] absorbing the impact on the terahertz link greatly limits the communication range of small base stations. This impact can be mitigated by the densification of the network.

Therefore, the terahertz can provide a rate of up to 16.4 Gbps with a delay threshold of 30 ms.

4.4.3 The application of THz technology in unmanned driving

At present, 5G has been put into use worldwide, and the B5G system will be a supplement to the current 5G. Due to the low latency and high reliability of THz technology, THz can be applied to driverless vehicles. The main goals of the current B5G system are as follows:

1. Extremely high data rates of each device (from tens of Gbps to Tbps)

2. A large number of connected devices

3. Ultra-large data rate per region

4. Ultra-reliable transmission, supporting various key applications, such as V2V communication, industrial control, and medical care

5. Part V: Prospect of V2X

5.1 Comparison of optical communication technology

The application of visible light, free optical communication and THz technology, and related technologies in V2X was described above. The following focuses on comparing related technologies.

5.1.1 Comparison of VLC and THz

Communication through visible light is a promising energy-sensing technology, attracting people from industry and academia to study its potential applications in different fields. VLC carries information by modulating light in the visible spectrum (390–750 nm) [45]. Recent advances in LED lighting have enabled unprecedented energy efficiency and lamp life because LEDs can be pulsed at very high speeds without significant impact on lighting output and the human eye. LED also has several attractive features, including low power consumption, small size, long life, low cost, and low heat radiation. Therefore, VLC can support many important services and applications, such as indoor positioning, human-computer interaction, device-to-device communication, vehicle networks, traffic lights, and advertising display [46]. Despite the advantages associated with deploying VLC communications, several challenges exist that may hinder the effectiveness of wireless communication links. To achieve high data rates in a VLC link, LoS channel should first be assumed, in which both the transmitter and receiver should be aligned with the field of view (FOV) to maximize channel gain. However, due to the continuous change of the movement and direction of the receiver, the field of view of the receiver may not always be aligned with the transmitter. This misalignment leads to a significant drop in received optical power [7]. When an object or man obstructs the line of sight, the optical power will drop significantly, resulting in a severe drop in the data rate. Similar to infrared waves, ambient light interference will significantly reduce the signal-to-noise ratio (SNR) of the received signal and reduce the communication quality [45]. The current research on visible light networks also reveals downstream traffic but does not consider how the uplink runs. Since the directional beam to the receiver should be maintained in the VLC uplink communication, when the mobile device is constantly moving/rotating, a significant throughput drop may occur. Therefore, other wireless technologies should be used to transmit uplink data [46]. Contrary to the VLC system, the THz band allows NLOS to propagate when LoS is not available as a supplement [47]. In this case, NLOS propagation can reflect the beam to the receiver by strategically installing dielectric mirrors. Due to the low reflection loss of the dielectric mirror, the resulting path loss is sufficient. In fact, for a distance of up to 1 meter and a transmission power of 1 watt, only the NLOS component in the terahertz link has a capacity of about 100 Gbps [48]. In addition, the terahertz frequency band is considered a candidate frequency band for uplink communication, which is a capability lacking in VLC communication. Another specific application where terahertz has become a valuable solution is the need to turn off the lights when looking for network services. Due to the limitation of the positive signal and the real signal, the VLC system will suffer a loss of spectrum efficiency. In fact, compared with the traditional bipolar system, using the unipolar OFDM system to impose Hermitian symmetry will cause a performance loss of 3 dB [49]. Both THz and VLC can be used as communication technologies to realize V2X in the future. Realize a technological breakthrough in V2X.

5.1.2 Comparison of VLC and FSO

VLC has become an attractive alternative to indoor RF communication to meet the growing demand for massive data services. In addition to providing a huge and unlicensed bandwidth to cope with the crowded radio spectrum, VLC has various other advantages, such as ease of use, no radiation, and no electromagnetic interference. On one hand, FSO is a line-of-sight, which has attracted great attention as a high-bandwidth last-mile transmission technology. On the other hand, FSO is a reasonable alternative to optical fiber because it requires less initial deployment [50] and it can be installed in locations where wired connection deployment is challenging.

The indoor VLC must be connected to the base station to achieve the purpose of communication. The most economical solution for connecting an indoor VLC to an outdoor base station is to use a power cord. In this case, various studies have been proposed [51, 52], involving the integration of VLC and power line communication (PLC) as a backbone network. However, PLC channels suffer from multiple damages—deep notches, high attenuation, and colored background noise that limits the data rate. To provide better data rates and improve system performance, VLC should be supplemented by high-bandwidth FSO links to achieve high-data-rate indoor multimedia services [53]. The direct FSO/VLC heterogeneous interconnection with data aggregation and distribution has been proved through experiments.

The combination of visible light and free optical communication technology can be used on V2X. Visible lights can be used on traffic signs, such as traffic lights. The combination of VLC and FSO can achieve V2I. The combination of optical communication and smart vehicles will provide better services. The combination of optical communication and smart vehicles will provide better services in the future smart vehicles.

5.1.3 Comparison of FSO and THz

FSO technology is an excellent candidate for high-performance secure communication due to safety, anti-interference, high beam directivity, flexibility, and energy efficiency. However, to date, the large-scale deployment of FSO communication systems has been affected by availability and reliability issues due to flicker on sunny days, low visibility on foggy days, Mie scattering effects, and high sensitivity to beam drift effects [54]. Due to the high directivity of the beam, FSO links are more difficult to intercept than RF systems. Nevertheless, Eve can still apply beam splitting attacks on the transmitting end, and blocking attacks or beam divergence attacks on the receiving end. Judging from the number of recent papers related to physical layer security (PLS). PLS research on FSO communication systems seems to be gaining momentum, such as [55, 56]. Unfortunately, almost all PLS papers related to FSO links use the eavesdropping channel method and direct detection introduced by Wyner [57]. Fog is the most unfavorable factor affecting FSO link reliability. In contrast, terahertz signals are less affected by these problems but are affected by other weather conditions, such as rain and snow. This shows that the two transmission media (FSO and THz) can operate in a complementary manner, depending on the prevailing weather and atmospheric conditions.

The above is the comparison of related technologies. In the future research direction, these several technologies can support the related research of V2X. The combination of pairwise or the combination of several technologies can play a very important role in future research.

5.2 Application scenarios of VLC

V-VLC is a fairly novel technology, although experimental studies in real driving scenarios have shown the feasibility of this technology in the application of vehicle networking. However, the current research on V-VLC mainly focuses on the understanding and characterization of the V-VLC channel, as well as the development of the V-VLC prototype. Although in the aspect of the physical layer, V-VLC still has many unsolved problems, especially regarding the performance and channel model of V-VLC in transportation system channels. But now the research will also focus on higher layer protocols V-VLC and IEEE 802.11p C-V2X and other different communication technologies that can make up for each other's shortcomings and improve the overall performance of applications [8].

Vehicular networking applications V-VLC can be used alone or implemented as part of a heterogeneous vehicular networking system V-VLC can benefit these specific applications as follows:

Cooperative sensing: The V-VLC can share perceptual data with nearby vehicles via onboard cameras, or collect sensor data to sense larger driving situations. Using headlights and taillights, high-throughput data can be transmitted to both front and rear vehicles, respectively, to facilitate cooperative awareness and cooperative awareness applications.

Information query: V-VLC can be used for information query and publishing within the range. These applications achieve this by utilizing VLC communication based on V2I and I2V to query and publish highly scalable and propagated information without strong latency and reliability requirements. In this way, V-VLC can be used to transmit information in part of the network without LED traffic lights, traffic signs, or road lighting coverage.

Intersection assistance: Intersection assistance applications, such as intersection collision avoidance, improve intersection safety by providing coordination and warning means between vehicles rather than traditional methods, such as traffic lights. When vehicles face each other at an intersection, head-to-head V-VLC links can be used to communicate with vehicles on the opposite side of the intersection. In addition, LED-based traffic lights or other infrastructure elements can facilitate communication.

Collision avoidance: To avoid rear-end collision in intelligent transportation systems, microwave radar and short-range radio communication are proposed.

However, these technologies are affected by radio frequency competition and changing weather conditions, and cannot achieve fully autonomous collision avoidance and safety queuing. VLC has the advantages of personal safety, unreasonable frequency allocation, large transmission capacity, and mature white LED light source. It can supplement the existing automatic driving system to achieve higher safety and driving efficiency, especially in the automotive lighting system and traffic light scenes [58].

Visible light localization system: Visible light system benefits from the ideal characteristics of visible light and its spectrum. Compared with traditional RF communication systems, visible spectrum has huge free bandwidth, facilitating high-speed data transmission and reducing the cost of operators. LED-based VLP systems can be easily integrated into existing lighting infrastructure (street light parking lights and traffic lights) for localized purposes, often without the need for rewiring beyond their basic lighting functions. In general, VLP systems can be used appropriately in any application that uses LEDs [59].

5.3 Applications of FSO in V2X

FSO can be used in future V2V systems by using highly collimated beams for enhanced vehicle-to-everything applications. A low-rate control link with multiple Gbps-assisted FSO links running in parallel is proposed [60]. Previous a control link is used to exchange sensor data about vehicle attitude dynamics to perform FSO beam tracking and provide an ultra-reliable high data rate connection on the latter FSO link. The joint contribution of local and distributed processing guarantees continuous and precise pointing to fully support autonomous driving applications.

To counteract the adverse effects of the limited sampling frequency of onboard sensors and control link delays, the evolution of vehicle kinematics can be estimated by simultaneously predicting and fusing multiple inertial measurement unit (IMU) data, augmented by real-time information on vehicle position.

5.4 Application of THz in V2X

When it comes to vehicle networks, there are several additional reasons to explore higher frequency bands that can support multiple Gbps and Tbps links. Firstly, when transmitting at such a high data rate, even if a user is mobile, from a data point of view, the link actually seems to be static because the transmission is almost instantaneous. In short, although the systems change over time, they do so much slower than the actual data rate. Therefore, during the transmission of a given frame, the system appears to be static. In addition, even if the user's connection is intermittent, the amount of information that can be transmitted per connection may be huge (1 Tb/ s). In addition, by moving to a higher carrier frequency, the influence of the Doppler effect can be reduced. Although this may not be a problem for automotive networks, it is very important for wireless data transmission between or between aircraft flying at high speeds. Therefore, there are inherent characteristics that prompt the exploration of vehicle networks in the terahertz frequency band [61]. This is undoubtedly the future trend for the realization of V2X. In the future, THz technology can shine in V2X.

Based on the results left over from the millimeter-wave band, the main attributes of terahertz communication are expected to include the following:

1. High frequency provides very large available bandwidth, therefore, potentially high data rate.

2. In response to high path loss, directional antennas will be mandatory. Highly directional antennas result in narrow beamwidths and very limited interference. Therefore, a very high data rate can be expected for each area.

3. If effective beam search and alignment mechanisms are in place, high rates can also lead to low delays.

THz will shine in unmanned driving. THz in V2X can be a direction and trend in the future, connecting everything. THz will deliver information quickly and efficiently in terms of high speed and high accuracy.

6. Part VI: Conclusion

V2X is the key technology of the Internet of Vehicles. The Internet of Vehicles in the true sense consists of the network platform, the vehicle, and the driving environment. This chapter focuses on the investigation and review of the important research results and forward-looking technologies of optical wireless communication in the application of V2X. VLC communication technology maximizes the use of existing traffic infrastructure to build a multi-user communication network structure for people-vehicle-traffic lights; the unique high-speed communication speed of THz and FSO communication technology provides strong communication speed support for V2X; IRS equipment research provides the possibility for long-distance NLOS communication. In addition, these aforementioned technologies and their key features are summarized, and their emerging future research and engineering directions are given. It is anticipated that, in building a smart city, optic-/THz-based technology will play an important role in a future highly developed V2X networking era.

Author details

Mingbo Niu, Xiaoqiong Huang* and Hucheng Wang
Vehicle-Road Collaborative Laboratory, Chang'an University, China

*Address all correspondence to: 2020232078@chd.edu.cn

IntechOpen

References

[1] Jurgen R. V2V, GPS integration could improve safety. In: V2V / V2I Communications for Improved Road Safety and Efficiency.

[2] Martín-Sacristán D, Roger S, Garcia-Roger D. Low-latency V2X communication through localized MBMS with local V2X servers coordination. In: 2018 IEEE International Symposium on Broadband Multimedia Systems and Broadcastin. 2018. pp. 1-8

[3] Huang J, Fei Z, Wang T, Wang X, Liu F, Haijun Z, et al. V2X-communication assisted interference minimization for automotive radars. China Communications. 2019;**16**(10):100-111

[4] Jurgen R. V2V and V2I Technical Papers. In: V2V/V2I Communications for Improved Road Safety and Efficiency.

[5] Malik RQ, Ramli KN, Kareem ZH, Habelalmatee MI, Abbas AH, Alamoody A. An overview on V2P communication system: Architecture and application. In: 2020 3rd International Conference on Engineering Technology and its Applications. 2020. pp. 174-178

[6] Matheus LEM, Vieira AB, Vieira LFM. Vieira MAM, Gnawali O. Visible light communication: Concepts, applications and challenges. IEEE Communications Surveys and Tutorials; 2019;**21**(4):3204-3237. DOI: 10.1109/COMST.2019.2913348

[7] Pathak PH, Feng X, Hu P, Mohapatra P. Visible light communication, networking, and sensing: A survey, potential and challenges. IEEE Communications Surveys and Tutorials; 2015;**17**(4):2047-2077. DOI: 10.1109/COMST.2015.2476474

[8] Memedi A, Dressler F. Vehicular visible light communications: A survey. IEEE Communications Surveys and Tutorials. 2021;**23**(1):161-181

[9] H. Burchardt, N. Serafimovski, D. Tsonev, S. Videv, and H. Haas, "VLC: Beyond point-to-point communication," IEEE Communication Magazine, vol. 52, no. 7, pp. 98–105, Jul. 2014.

[10] Rohner C, Raza S, Puccinelli D, Voigt T. Security in visible light communication: Novel challenges and opportunities. IEEE Sensors Transducers. 2015;**192**(9):9-15

[11] Hansen CJ. WiGiG: Multi-gigabit wireless communications in the 60 GHz band. IEEE Wireless Communication. 2011;**18**(6):6-7

[12] Gomez A, Shi K, Quintana C. Beyond 100-Gb/s indoor wide field-of-view optical wireless communications. IEEE Photonics Technology Letters. 2015;**27**(4):367-370

[13] Yeh CH, Chow CW, Wei LY. 1250 Mbit/s OOK wireless white-light VLC transmission based on phosphor laser diode. IEEE Photonics Journal. 2019; **11**(3):1-5

[14] Zohaib A, Ahfay MH, Mather PJ, Sibley MJN. Improved BER for offset pulse position modulation using priority decoding over VLC system. In: IEEE Wireless Days, Manchester, UK. 2019

[15] Sui M, Zhou Z. The modified PPM modulation for underwater wireless optical communication. In: 2009 International Conference on Communication Software and Networks, Chengdu, China. 2009

[16] Islam MA, Chowdhury AB, Barua B. Free space optical communication with m-ary pulse position modulation under strong turbulence with different type of

receivers. In: 2015 2nd International Conference on Electrical Information and Communication Technologies, Khulna, Bangladesh. 2015

[17] Chizari A, Jamali MV, AbdollahRamezani S, Salehi JA, Dargahi A. Designing a dimmable OPPM-based VLC system under channel constraints. In: 2016 10th International Symposium on Communication Systems, Networks and Digital Signal Processing, Prague, Czech Republic. 2016

[18] Armstrong J. OFDM for optical communications. IEEE Journal of Lightwave Technology. 2009;**27**(3): 189-204

[19] Li F, Zhang C. Performance Analysis on Hybrid OOK-and-OFDM Modulation in VLC System. In: 2019 IEEE International Symposium on Broadband Multimedia Systems and Broadcasting, Jeju, Korea. 2019

[20] Das S, Henniger H, Epple B. Requirements and challenges for tactical free-space lasercomm[C]. In: MILCOM 2008-2008 IEEE Military Communications Conference. 2008. pp. 1-10

[21] Andrews LC, Phillips RL, Hopen CY, AI-Habash MA. Theory of optical scintillation. Optical Society of America. 1999;**16**:1417-1429

[22] Haas H. LiFi: Conceptions, misconceptions and opportunities. In: 2016 IEEE Photonics Conference. IEEE. 2016

[23] Ravishankar MB, Rakshith RA, Aishwarya A, Thejaswini KN, Basavaraja G. Free space optics and radio frequency signals in spacial communication. In: 2019 International Conference on Intelligent Computing and Control Systems. 2019. pp. 1473-1477

[24] Chen S, Zhang Z, Cai S. High power budget coherent free space optical communication system based on fiber laser phased array. In: 2018 Asia Communications and Photonics Conference. 2018. pp. 1-3

[25] Kumar N, Rana AK. Impact of various parameters on the performance pf free space optics communication system. Optik-International Journal for Light and Electron Optics. 2013;**124**: 5774-5776

[26] Zocchi FE. A simple analytical model of adaptive optics for direct detection free space optical communication. Optics Communications. 2005;**248**:359-374

[27] Hamza AS, Deogun JS, Alexander DR. Free space optical multicast crossbar. IEEE/OSA Journal of Optical Communications and Networking. 2016;**8**(1):1-10

[28] Jingzhi Hu, Hongliang Zhang, Boya Di, Lianlin Li, Kaigui Bian, Lingyang Song, and Yonghui Li, "Reconfigurable intelligent surface based RF sensing: Design, optimization, and implementation," in IEEE Journal on Selected Areas in Communications, vol. 38, no. 11, pp. 2700-2716, Nov. 2020.

[29] Index CVN. Cisco visual networking index: Forecast and methodology 2015–2020. In: White paper, CISCO. 2015

[30] Li R. Towards a new Internet for the year 2030 and beyond. In: 3rd Annual ITU IMT-2020/5G Workshop Demo Day, Geneva, Switzerland. 2018. pp. 1-21

[31] Kerecman AJ. The tungsten-p type silicon point contact diode. In: MTT-S IEEE Int. Microw. Symp. Dig. 1973. pp. 30-34

[32] Fleming JW. High resolution submillimeter-wave Fouriertransform spectrometry of gases. IEEE

Transactions on Microwave Theory and Techniques. 1974;**MTT-22**(12): 1023-1025

[33] Siegel PH. Terahertz technology. IEEE Transactions on Microwave Theory and Techniques. 2002;**50**(3): 910-928

[34] Ferguson B, Zhang X-C. Materials for terahertz science and technology. Nature Materials. 2002;**1**(1):26

[35] Piesiewicz R, Kleine-Ostmann T, Krumbholz N, Mittleman D, Koch M. Short-range ultra-broadband terahertz communications: Concepts and perspectives. IEEE Antennas Propagation Magazine. 2007;**49**(6): 24-39

[36] Nagatsuma T. A 120-GHz integrated photonic transmitter. In: Proc. IEEE Int. Topical Meeting Microw. Photon. 2000. pp. 225-228

[37] Hirata A, Kosugi T, Takahashi H. 120-GHz-band wireless link technologies for outdoor 10-Gbit/s data transmission. IEEE Transactions on Microwave Theory and Techniques. 2012;**60**(3):881-895

[38] Takahashi H, Hirata A, Takeuchi J, Kukutsu N, Kosugi T, Murata K. 120-GHz-band 20-Gbit/s transmitter and receiver MMICs using quadrature phase shift keying. In: Proc. IEEE 7th Eur. Microw. Integr. Circuits Conf. 2012. pp. 313-316

[39] Akyildiz IF, Jornet JM, Han C. TeraNets: Ultra-broadband communication networks in the terahertz band. IEEE Communication Magazine. 2014;**21**(4):130-135

[40] Larsson EG, Edfors O, Tufvesson F, Marzetta TL. Massive MIMO for next generation wireless systems. IEEE Communication Magazine. 2014;**52**(2): 186-195

[41] Akyildiz IF, Jornet JM. Realizing ultra-massive MIMO communication in the terahertz band. Nano Communication Network. 2016;**8**:46-54

[42] Zakrajsek LM, Pados DA, Jornet JM. Design and performance analysis of ultra-massive multi-carrier multiple input multiple output communications in the terahertz band. In: Proc. SPIE Image Sens. Technol. Mater. Devices Syst. Appl. IV. Vol. 10209. 2017. pp. 102090A1-102090A11

[43] Muñoz SR. Multi-user ultra-massive MIMO for very high frequency bands (mmWave and THz): A resource allocation problem [M.S. thesis]. Barcelona, Spain: Dept. Comput. Architect, Universitat Politècnica de Catalunya; 2018

[44] Chaccour C, Amer R, Zhou B, Saad W. "On the reliability of wireless virtual reality at terahertz (THz) frequencies". arXiv preprint arXiv: 1905.07656, 2019

[45] Arnon S. Visible Light Communication. Cambridge, UK: Cambridge Univ. Press; 2015

[46] Khalighi MA, Uysal M. Survey on free space optical communication: A communication theory perspective. IEEE Communication Surveys Tuts. 2014;**16**(4):2231-2258

[47] Akyildiz IF, Jornet JM, Han C. Terahertz band: Next frontier for wireless communications. Physics Communication. 2014;**12**:16-32

[48] Moldovan A, Ruder MA, Akyildiz IF, Gerstacker WH. LOS and NLOS channel modeling for terahertz wireless communication with scattered rays. In: Proc. IEEE GC Wkshps. 2014. pp. 388-392

[49] Wang Z, Mao T, Wang Q. Optical OFDM for visible light communications. In: Proc. IEEE 13th Int. Wireless

Commun. Mobile Comput. Conf. 2017. pp. 1190-1194

[50] Douik A, Dahrouj H, Al-Naffouri TY, Alouini M-S. Hybrid radio/free-space optical design for next generation backhaul systems. IEEE Transactions on Communications. 2016;**64**(6):2563-2577

[51] Komine T, Nakagawa M. Integrated system of white LED visiblelight communication and power-line communication. IEEE Transactions on Consume Electronics. 2003;**49**(1):71-79

[52] Ma X, Gao J, Yang F, Ding W, Yang H, Song J. Integrated power line and visible light communication system compatible with multi-service transmission. IET Communication. 2017;**11**(1):104-111

[53] Huang Z, Wang Z, Huang M, Li W, Lin T, He P, et al. Hybrid optical wireless network for future SAGO-integrated communication based on FSO/VLC heterogeneous interconnection. IEEE Photonics Journal. 2017;**9**(2):1-10

[54] Andrews LC, Philips RL. Laser Beam Propagation through Random Media. Bellingham, WA: SPIE Press; 2005

[55] Sun X, Djordjevic IB. Physical-layer security in orbital angular momentum multiplexing free-space optical communications. IEEE Photonics Journal. 2016;**8**(1):1-10

[56] Lopez-Martinez FJ, Gomez G, Garrido-Balsells JM. Physical layer security in free-space optical communications. IEEE Photonics Journal. 2015;**7**(2):7901014

[57] Wyner AD. The wire-tap channel. Bell System Technology Journal. 1975;**54**(8):1355-1387

[58] Soner B, Coleri S. Visible light communication based vehicle localization for collision avoidance and platooning. IEEE Transactions on Vehicular Technology. 2021;**70**(3): 2167-2180

[59] Keskin MF, Sezer AD, Gezici S. Localization via Visible Light Systems. Proceedings of the IEEE. 2018;**106**(6): 1063-1088

[60] Brambilla M, Tagliaferri D, Nicoli M, Spagnolini U. Sensor and Map-Aided Cooperative Beam Tracking for Optical V2V Communications. In: 2020 IEEE 91st Vehicular Technology Conference. 2020. pp. 1-7

[61] Shahid M, Miquel JJ, Jocelyn A, Gerstacker Wolfgang H, Xiaodai D, Bo A. Terahertz communication for vehicular networks. In: IEEE Transactions on Vehicular Technology. 2017

Prediction of Large Scale Spatio-temporal Traffic Flow Data with New Graph Convolution Model

Ping Wang, Tongtong Shi, Rui He and Wubei Yuan

Abstract

Prompt and accurate prediction of traffic flow is quite useful. It will help traffic administrator to analyze the road occupancy status and formulate dynamic and flexible traffic control in advance to improve the road capacity. It can also provide more precise navigation guidance for the road users in future. However, it is hard to predict spatio-temporal traffic flow data in large scale promptly with high accuracy caused by complex interrelation and nonlinear dynamic nature. With development of deep learning and other technologies, many prediction networks could predict traffic flow with accumulated historical data in time series. In consideration of the regional characteristics of traffic flow, the emerging Graph Convolutional Network (GCN) model is systematically introduced with representative applications. Those successful applications provide a possible way to contribute fast and proper traffic control strategies that could relieve traffic pressure, reduce potential conflict, fasten emergency response, etc.

Keywords: traffic flow, GCN, traffic data, deep learning, ITS

1. Introduction

1.1 Background and current status

Traffic problems such as frequent traffic congestion, serious traffic accidents, and long commuting times have seriously reduced the travel experience of passengers and the efficiency of traffic operations [1]. To cope with these problems, researchers work on improving the traffic control strategies based on prediction of future traffic stratus [2]. Traffic flow is one of important road conditions to access [3]. Based on prompt and accuracy perdition, better and fast-adjusted traffic control and guidance could be applied. Therefore, reliable traffic flow prediction is also one of the key factors to upgrade the traffic system from "passive adjust" to "active control in advance"; even prediction of future short-term traffic status of road sections is quite useful to prevent congestion deteriorate. For traffic management departments, early detection of traffic instability and abnormal potential risks based on reliable prediction data can improve a large number of existing traffic management control applications, such as traffic calming, signal control, etc.; for road users, real-time route updates and adjustments based on dynamic traffic prediction results can adjust travel time and routes before congestion develops, thus providing vehicles to plan a driving path to avoid congested road sections and congested intersections or to plan a path with the shortest driving time for vehicles to improve traffic efficiency.

The current traffic prediction also faces the following challenges, as shown in **Figure 1**: (1) analyzing the spatial correlation of the road network: some roads are adjacent to each other and have different degrees of influence on upstream and downstream traffic volumes, so the traffic flow in this part is spatially correlated, and it is a challenge to consider the spatial location relationship to correlate the neighboring traffic flow characteristics [4]; (2) unlike the regular network layout, the traffic map structure is irregular; (3) the nonlinear retention of medium and long time prediction models: the traffic flow changes drastically at the peak time, which is difficult to predict, especially as the prediction time increases, the nonlinear retention ability of the model decreases and the time series signal gradually decays, so how to better correlate the time series relationship of traffic flow to maintain the steady-state time series prediction is also a long-term challenging task [5].

1.2 Related works

Traffic flow forecasting is based on historical traffic flow data to predict future traffic flow, which is a typical regression problem of traffic network time series [6]. In order to solve the traffic flow forecasting problem, factors such as traffic patterns, data types, spatial locations, and time periods need to be considered. Nowadays, many computational forecasting methods have been widely used in traffic flow prediction and have achieved good research results. As shown in **Figure 2**, common traffic flow prediction methods can be divided into three major categories [7]. ① early traffic flow prediction methods; ② machine-learning-based traffic flow prediction methods; ③ deep-learning-based traffic flow prediction methods.

1.2.1 Early traffic flow prediction methods

Early traffic flow prediction methods mainly model the relationship between traffic flow, speed, and density and regress the traffic flow data as well as optimize the parameters to achieve the fitting prediction of traffic data, mainly including statistical models and traffic simulation.

1.2.1.1 Typical methods

- Miska et al. [8] proposed cellular automata (CA) to simulate each participant of different flows and their interaction phenomena.

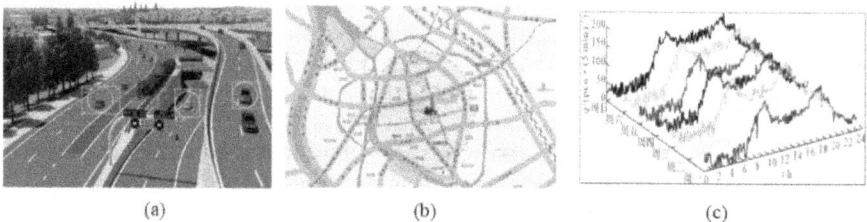

(a)   (b)   (c)

Figure 1.
Challenges in traffic flow prediction; (a) road relevance from the web (https://www.ivsky.com/tupian/daolu_t948/); (b) the complex road network map of Shaanxi Province; (c) periodicity of traffic data.

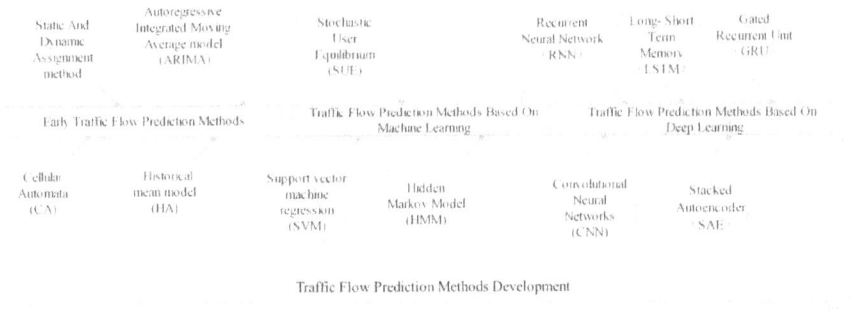

Figure 2.
Classification of time-line based traffic flow prediction methods.

- Ngoduy et al. [9] proposed the use of static and dynamic assignment methods to allocate traffic on a simulated road network.

- Stephanedes et al. [10] have applied the historical mean model (HA) in urban traffic control systems in 1984.

- Kumer et al. [11] used the Autoregressive Integrated Moving Average (ARIMA) model to represent the predicted traffic flow in the form of a mathematical model.

1.2.1.2 Advantages and disadvantages

Models such as statistical mathematical models and traffic simulations can describe this traffic flow prediction as a time series problem approximately. However, simulation systems and simulation tools still need to consume a lot of computational power and skilled parameter settings to reach a steady state, and it is more difficult to get accurate prediction results from this prediction model due to the complexity of traffic scenarios. Besides, these methods based on statistics are only applicable to linear data, while traffic flow data are nonlinear and complex; thus, such methods are not capable of handling complex nonlinear traffic data.

1.2.2 Traffic flow prediction methods based on machine learning

With the demand for high accuracy in intelligent traffic scenarios, the shortcomings of traditional prediction methods that cannot model the complex state of traffic flow become more and more prominent, and machine learning methods gradually take an important place in traffic flow prediction tasks.

1.2.2.1 Typical methods

- YS et al. [12] proposed a traffic flow prediction method using support vector machine regression. The idea of using support vector machine method is to map the low-dimensional nonlinear traffic data to a high-dimensional space by introducing a kernel function before linear classification.

- Zhu et al. [13] predict the path traffic volume and roadway flow by building a Bayesian network model.

- Qi et al. [14] proposed a Hidden Markov Model (HMM) for short-term highway traffic prediction.

1.2.2.2 Advantages and disadvantages

Machine learning methods can better model the stochastic processes and nonlinear properties of traffic flows and have mostly better performance compared with traditional models. However, such methods do not consider the spatial and temporal correlation of traffic flow data and require extensive feature engineering. Therefore, it is difficult to solve complex traffic flow prediction problems.

1.2.3 Traffic flow prediction methods based on deep learning

Deep learning has been very successful in the fields of computer vision, speech recognition, and natural language processing, and more and more scholars are applying deep neural networks (DNNs) to various real-world scenario tasks. In traffic flow prediction, the models can be classified into road section prediction and area prediction according to their prediction range characteristics.

1.2.3.1 Typical methods

- Chen et al. [15] proposed a convolutional neural network (CNN)-based traffic flow prediction method using time series folding for multi-scale learning.

- Lv et al. [16] proposed a heap-based autoencoder (SAE) method for traffic flow prediction considering spatiotemporal relationships.

- Yu et al. [17] proposed a Long-Short Term Memory (LSTM)-based method for traffic flow prediction on road networks during peak periods.

- Cho K et al. [18] proposed a gated recurrent unit that can establish links for traffic data at adjacent moments and preserve the memory by gating and other means to learn long-term dependencies of traffic flow sequences.

- Yu et al. [19] proposed a spatiotemporal graph convolutional network (STGCN) for traffic flow velocity prediction on a multi-scale traffic network.

- The ST-ResNet [20] model restricts the input to grid data rather than graph structure in the traffic prediction problem, which makes it difficult to make predictions on complex highway data.

- Geom-GCN [21] proposed a network to update the node representation, but it could not capture the distance dependence between nodes.

- The DCRNN model (2018ICLR) [22] models spatial correlation as a diffusion process on directed graphs to model the translation of traffic flows and proposes diffusion convolutional recurrent neural networks capable of capturing spatial and temporal dependencies between time series using the seq2seq framework.

- The GMAN model (2020AAAI) [23] uses an attention mechanism to model dynamic spatial and nonlinear temporal relationships, respectively.

ASTGCN [24] considers only low-order neighborhood relationships between nodes and ignores correlations between different historical time periods.

1.2.3.2 Advantages and disadvantages

The core of the traffic prediction problem lies in how to effectively capture the spatiotemporal dimensional features and correlations of the data. Traditional convolutional neural networks can effectively extract local features of data, but can only work on standard grid data. The graph convolution can directly extract features from graph structured data and automatically mine the spatial patterns of traffic data. The convolution operation along the time axis can extract the temporal patterns of traffic data. Therefore, this paper focuses on the deep learning model based on graph convolutional network to capture the spatial and temporal characteristics of traffic data and effectively solve the traffic flow prediction problem.

1.2.4 Future directions for exploring traffic flow prediction models

GCN has become a mainstream method in the field of traffic flow prediction, but it started late, and its theoretical foundation and research depth are far from enough. At present, it still faces many problems that need to be solved. There are three main directions as follows.

1. Dynamic graph modeling: Most graph structures processed by GCN networks are static graphs, and there are fewer models involving dynamic graph structures. The graph structure of static graphs is static and unchanging, while the vertices and edges of dynamic graphs change randomly or even disappear, making it difficult to follow the rules.

2. Heterogeneous graph modeling: Homogeneous graph means that nodes and edges are only one type, and this kind of data is easier to handle. The heterogeneous graph refers to the type of nodes and edges, the same node and different node connections will show different properties, the same edge and different node connections will also show different relationships, heterogeneous graph structure is relatively complex to deal with. However, the heterogeneous graph is the most relevant scenario to the actual problem.

3. Deepening the model structure: One of the inherent advantages of GCN is that it smoothes the graph signal, but as its layers keep deepening, its training results are highly susceptible to over smoothing. Since graph convolution is a special form of Laplacian smoothing, the smoothing operation makes the signal more consistent at the feature level as the graph convolution aggregates the features of neighboring nodes, thereby causing the signal to lose its diversity and leading to a sharp performance degradation in the relevant prediction task, a phenomenon that is more pronounced on small data sets. Therefore, GCNs cannot be stacked continuously and deeply like general convolutional models, but shallow neural networks suffer from limited perceptual field and feature extraction capabilities.

2. Traffic prediction based on graph convolution

This section introduces the principles and techniques related to traffic flow prediction based on graph neural networks. First, an overview of graph neural networks is given, and the graph convolutional networks (GCNs) [25] used to capture the spatial dependence of traffic flows in the road network are introduced separately in this paper. Secondly, the transformation of graph structure into actual road traffic graph structure modeling method is introduced; finally, this paper models the GCN on urban road networks and uses the topology of the GCN capture graph to handle the spatiotemporal traffic prediction task, and the application scenarios of traffic flow prediction are added at the end of the paper.

2.1 Basic graph theory and convolutional networks

2.1.1 Graph theory

Graph is a common data structure that is an important object of study in the field of computer and data science [26]. A graph usually consists of two elements, Vertex and Edge, where the vertices correspond to an abstract representation of the object of study and the edges represent the interconnection between two of the objects. Graphs are often used to represent things and specific relationships between things; in fact, graphs can represent any system with binary relationships. Graphs have a very wide range of applications in real life; social networks of human life, citation systems, urban transportation networks, and biochemical molecules can be effectively represented by graph structures.

2.1.1.1 Basic concept

In graph theory, a graph is usually represented as a set of vertices and edges [27], denoted as $G = (V, E)$, where $V = \{v_1, v_2, \ldots, v_n\}$ is the set of vertices and the elements in this non-empty set are the vertices of the graph. The set of edges can be denoted as E. A graph can be classified into directed and undirected graphs depending on whether the edges in the graph have directionality or not. If the edges of a graph have directionality, such edges are called directed edges as in **Figure 3(a)**, and if the edges of a graph have no directionality, then the corresponding graph is an undirected graph as in **Figure 3 (b)**. The graphs can be classified into weighted and unweighted graphs according to the presence or absence of specific weights of the edges in the graph [28]. Each edge in a weighted graph has a real weight as in **Figure 3(c)**, which represents the degree of connection between two vertices or the "distance" between two vertices. For example, in a traffic network, the weight of an edge can characterizes the physical distance between two vertices. In contrast, the other category is the unweighted graph, which can also be understood as the unweighted graph in which all the edge weights are equal.

2.1.1.2 Algebraic representation of graphs

As a common data structure, graphs have many kinds of algebraic representations, and common storage representations include adjacency matrices [29],

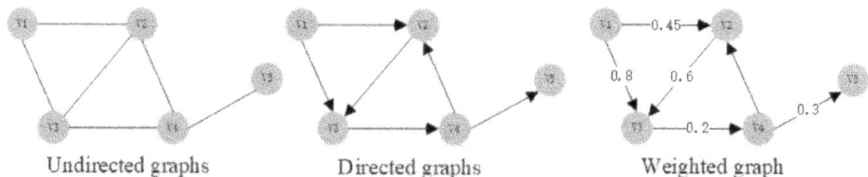

| Undirected graphs | Directed graphs | Weighted graph |

Figure 3.
Basic types of common diagrams.

adjacency tables, and association matrices. Among them, adjacency matrices are widely used in graph representation learning because they can represent the constructional properties of graphs well and are easy to combine with matrix operations to understand the structural features of graphs.

If two vertices of an edge in a graph are v_i and v_j, then v_i and v_j are said to be their respective neighbors. We define the set of neighbors of v_i as $N(v_i)$:

$$N(v_i) = \{v_j | \exists e_{ij} \in E \text{ or } e_{ji} \in E\} \qquad (1)$$

The degree of v_i is defined as: the number of edges with v_i as the endpoint, denoted as $deg(v_i)$, and therefore, $deg(v_i) = N(v_i)$. In a directed graph, the degree of a vertex can be divided into out-degree and in-degree. The number of directed edges starting at vertex v_i is called the out-degree of v_i, and the number of directed edges ending at vertex v_i is called the in-degree of v_i. The sum of the entry and exit degrees of a vertex is equal to the degree of the vertex, and the sum of the degrees of all nodes is equal to twice the number of all edges. e_{ij} and e_{ji} represent edges in different directions between two vertices, e.g., e_{ij} represents an edge in the direction from i to j, while e_{ji} is the opposite.

The degree matrix is a matrix of the degrees of the vertices, so that the elements at the main diagonal positions are the vertex degrees and the remaining elements are 0. Accordingly, the directed graph has an entry degree matrix and an exit degree matrix. The adjacency matrix is a matrix used to represent the relationship between vertices. For graph $G = (V, E)$, the adjacency matrix can be expressed as:

$$A_{ij} = \begin{cases} 1 & \text{if } <v_i, v_j> \subseteq E \\ 0 & \text{else} \end{cases} \qquad (2)$$

The core idea of the adjacency table of a graph is to have a neighbor table for each vertex of the vertex set. The association matrix is used to represent the direct association of nodes and edges and is defined as:

$$B_{ij} = \begin{cases} 1 & \text{if } v_i \text{ and } e_j \text{ are connected} \\ 0 & \text{else} \end{cases} \qquad (3)$$

The Laplace matrix [30] is a special matrix that is often used in graph theory to study the structural properties of graphs. The Laplace matrix is defined as $L = D - A$, where D is the degree matrix of the graph and A is the adjacency matrix of the graph. **Figure 4** shows the Laplace matrix representation of a simple graph.

2.1.2 Graph convolutional networks

Previous classical convolutional networks based on deep learning mostly consider regular data in Euclidean space in processing data. When inputting ordered data with fixed dimensions (e.g., images, speech, video, etc.), the convolutional operation and the capture and compression of the pooling layer make the network fitting effect remarkable. However, when faced with sequentially disordered road network traffic data with variable dimensions, the suitability of the traditional convolution operation decreases. However, graph neural networks (GNNs) can handle the abovementioned irregular graphs by passing node features into the neural network during iteration and outputting the node states. The original GNNs converge the hidden state to a fixed point based on the "immobile point" theory, which is ineffective for extracting edge information, and in the specific scenario

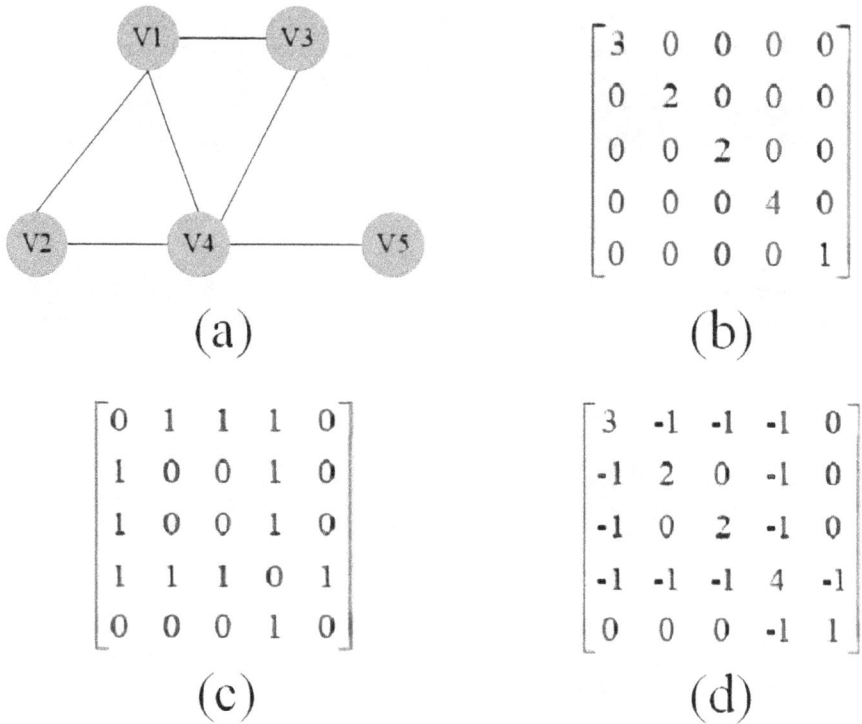

$$\begin{bmatrix} 3 & 0 & 0 & 0 & 0 \\ 0 & 2 & 0 & 0 & 0 \\ 0 & 0 & 2 & 0 & 0 \\ 0 & 0 & 0 & 4 & 0 \\ 0 & 0 & 0 & 0 & 1 \end{bmatrix}$$

(a)   (b)

$$\begin{bmatrix} 0 & 1 & 1 & 1 & 0 \\ 1 & 0 & 0 & 1 & 0 \\ 1 & 0 & 0 & 1 & 0 \\ 1 & 1 & 1 & 0 & 1 \\ 0 & 0 & 0 & 1 & 0 \end{bmatrix} \qquad \begin{bmatrix} 3 & -1 & -1 & -1 & 0 \\ -1 & 2 & 0 & -1 & 0 \\ -1 & 0 & 2 & -1 & 0 \\ -1 & -1 & -1 & 4 & -1 \\ 0 & 0 & 0 & -1 & 1 \end{bmatrix}$$

(c)   (d)

Figure 4.
Matrix representation of the graph; (a) graph structure; (b) degree matrix; (c) adjacency matrix; (d) Laplace matrix.

represented by the graph, some feature information is shared among nodes due to the fixed convergence, making the actual information obtained scarce. Therefore, two types of Graph Convolutional Network (GCN) based on frequency domain and null domain are generated. Two types of GCN models: null domain convolution is the same as the traditional convolution method, which can convolve directly at the pixel point of the picture; frequency domain convolution needs to start from the graph signal processing, treating the kernel in the convolution as a filter and the learned features as signals for weighted summation.

As shown in **Figure 5**, the common network framework for graph convolution is illustrated. First, the neighboring nodes of the input graph structure are updated with a layer of convolution operation, and then a layer of ReLU activation function is added to obtain the basic convolution layer plus activation function structure. The above structure is stacked sequentially until the number of stacked layers reaches the prediction of the model, and the output part transforms the node features into labels for the relevant tasks. Unlike GNN circular iterative parameter sharing, GCN is a multilayer stack and the parameters are different for each layer.

Further, it mainly includes graph convolution based on the spectral domain (frequency domain) and the null domain. The spectral domain approach is to construct CNN simulations into the spectral domain by considering the localization of graph convolution through spectral analysis, such as Spectral Graph Convolution (SGC), which mainly focuses on the continuous derivation and improvement of the core formulations of spectral graph theory to reduce the computational power of the model from the perspective of optimization parameters. Empty domain

| Input | 1 layer convolution | ReLU | 1 layer convolution | ReLU | Circular Stacking | Output Node Tasks |

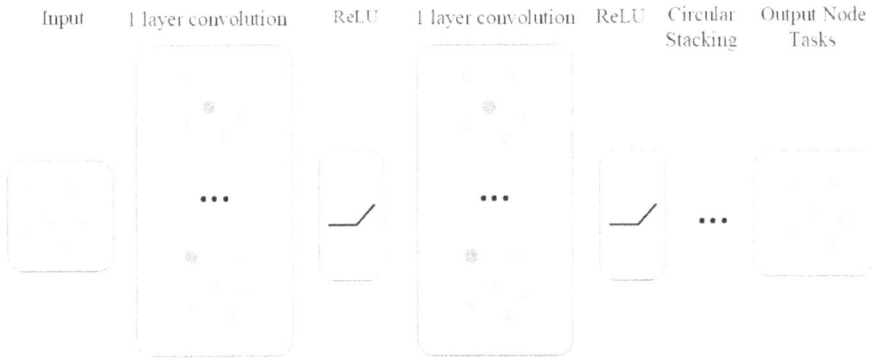

Figure 5.
General framework of graph convolution.

methods perform convolution filters directly on the nodes of the graph and their neighborhoods, such as Diffusion Graph Convolution (DGC).

2.1.2.1 Spectral Domain-based Graph Convolution Network (SGC)

Spectral domain approach [31]: The absence of graph translation invariance poses difficulties in defining convolutional neural networks in the nodal domain. The spectral domain approach uses the convolution theorem to define the graph convolution from the spectral domain. The spectral domain graph convolution network is proposed based on graph signal processing, where the convolution layer of the graph neural network is defined as a filter, i.e., the filter removes the noise signal to obtain the result of the input signal. In practical applications, it can only be used to process graph structures that are undirected and have no information on the edges. The Fourier transform of the signal f(x) and its inverse transform are:

$$F(w) = \varphi(w) = \int_{-\infty}^{+\infty} f(x) \exp(-iwx) dx \qquad (4)$$

$$f(x) = \varphi^{-1}(F(w)) \qquad (5)$$

where φ varphi denotes the Fourier transform. It can be found that the Fourier transform that changes the time domain to the spectral domain is essentially the integral of the summation $f(x)$ with $\exp(-iwx)$ as the basis vector. Defining the graph G of the input signal as a characteristic decomposable Laplace matrix $L = D - A$, the normalized Laplace matrix L is defined as:

$$L = I_n - D^{-\frac{1}{2}} A D^{-\frac{1}{2}} \qquad (6)$$

where D denotes the degree matrix of graph G, A denotes the adjacency matrix, and I_n is the unit matrix of order n. After performing the eigendecomposition, it can be expressed as the universal structure $L = U\Lambda U^T$. where Λ is a matrix with each eigenvalue as a diagonal element, and U is a vector matrix composed of eigenvectors corresponding to each eigenvalue. Since U is an orthogonal matrix, the basis of the conventional Fourier transform $\exp(-iwx)$ is then replaced by U^T and expressed in matrix form to obtain the Fourier transform of the signal x on the graph as:

$$\hat{x} = U^T x \qquad (7)$$

where x refers to the original representation of the signal, $\hat{x}$ refers to the signal x after transforming it to the spectral domain, and U^T denotes the transpose of the eigenvector matrix for doing the Fourier transform. The inverse Fourier transform of the signal x is:

$$x = U\hat{x} \tag{8}$$

Using the Fourier transform and the inverse transform on the graph, the graph convolution operation can be implemented as follows.

$$x \overset{*}{_G} g = U\left((U^T x) \odot (U^T y)\right) \tag{9}$$

where $\overset{*}{_G}$ as denotes the graph convolution operator, x denotes the signal in the node domain on the graph, g is the graph convolution kernel, and $\odot$ refers to the Hadamard product, which denotes the multiplication of the corresponding elements of two vectors. By replacing the vector $U^T y$ with the diagonal array g_θ theta, the Hadamard product is transformed into a matrix multiplication. The graph convolution operation is denoted as $U g_\theta U^T x$.

To solve the excessive computation of Laplace eigenvalues and eigenvectors, Defferrar et al. [32] proposed ChebNet based on Chebyshev polynomials. The eigenvalue matrix is approximated by Chebyshev polynomials, and the Chebyshev polynomials are as follows.

$$g_\theta(\Lambda) = \sum_{k=0}^{k-1} \theta_k T_k(\tilde{\Lambda}) \tag{10}$$

where θ is the Chebyshev coefficient; $T_k(\tilde{\Lambda})$ is the k-th order Chebyshev polynomial of $\tilde{\Lambda}$; $\tilde{\Lambda} = \frac{2\Lambda}{\lambda_{max}} - I_n$, $\tilde{\Lambda}$ are the normalized eigenvalue diagonal matrices. Thus, the convolution operation can be expressed as follows:

$$x \overset{*}{_G} g_\theta = U\left(\sum_{k=0}^{k-1} \theta_k T_k(\tilde{\Lambda})\right) U^T x = \sum_{k=0}^{k-1} \theta_k T_k(\tilde{L}) x \tag{11}$$

where $\frac{2L}{\lambda_{max}} - I_n$; the computational complexity of the graph convolution calculation is reduced from $O(N^2)$ to $O(LE)$ by replacing the Chebyshev expansion $T_k(\tilde{\Lambda})$ with the eigen-decomposition part of the frequency domain convolution g_θ in the original GCN, effectively avoiding the computational part of the eigen-decomposition, where E is the number of edges in the input graph and L is the order of the Laplace operator polynomial. ChebNet results in a significant reduction in computational complexity and a significant improvement in computational efficiency.

After that, Kipf et al. used first-order Chebyshev polynomials and simplified the spectral graph convolution by restricting the parameters in order to make ChebNet have better local connectivity properties. Let $K = 2$, $T_0(\tilde{L}) = 1$, $T_1(\tilde{L}) = L$, $\lambda_{max} = 2$. Then the graph convolution calculation is simplified as:

$$x \overset{*}{_G} g_\theta \approx \theta_0 x + \theta_1(L - I_n)x = \theta_0 x - \theta_1 D^{-\frac{1}{2}} A D^{-\frac{1}{2}} x \tag{12}$$

Where: θ_0 and θ_1 are free parameters, shared by the whole graph. Let $\theta = \theta_0 = -\theta_1$, i.e., the two parameters are transformed into a one-parameter model, then the graph convolution is calculated as:

$$x_G^* g_\theta \approx \theta \left(I_n + D^{-\frac{1}{2}} A D^{-\frac{1}{2}} \right) x \tag{13}$$

However, since $I_n + D^{-\frac{1}{2}} A D^{-\frac{1}{2}}$ is the eigenvalue of $[0, 2]$, which may lead to the problem of disappearing, exploding or unstable values of neural network gradients, $\tilde{D}^{-\frac{1}{2}} \tilde{A} \tilde{D}^{-\frac{1}{2}}$ is used instead of $I_n + D^{-\frac{1}{2}} A D^{-\frac{1}{2}}$ for normalization.

2.1.2.2 Graph Convolutional Network (DGC) based on spatial domain

The spatial domain approach: spatial-based graph convolutional networks were first proposed in Neural Network for Graphs (NN4G), which is different from the spectral domain graph convolutional neural network from signal processing theory, the spatial domain graph convolutional neural network starts from the nodes in the graph, designs the aggregation function to gather the features of neighboring nodes, adopts the message propagation mechanism, and thinks about how to accurately and efficiently use the features of neighboring nodes of the central node to update the features of the central node. The essence of CNN is weighted summation, and the spatial domain graph convolutional neural network is based on the basic construction process of CNN to accomplish the purpose of GNN aggregation of neighboring nodes from the perspective of summation. Since the nodes in the graph are unordered and the number of neighboring nodes is uncertain, one idea of the spatial domain graph convolutional neural network is (1) to fix the number of neighboring nodes and (2) to sort the neighboring nodes. If the above two tasks are completed, the non-Euclidean structured data becomes ordinary Euclidean structured data, and naturally the traditional algorithm can be completely migrated to the graph. Among them, step (1) also facilitates the application of GNN to graphs with many nodes.

Currently, GCN has become a fundamental model for traffic flow prediction research and a benchmark method for experiments. Although neither the air-domain graph convolution network nor the frequency-domain graph convolution network is proposed for the traffic flow prediction problem, the natural graph structure property of traffic data makes GCN show high efficiency and accuracy in the field of traffic flow prediction than the traditional methods.

2.2 Modeling experiments for traffic prediction

This section will first give a specific definition of the traffic flow prediction problem and then give the flow of the traffic flow prediction model based on spatiotemporal characteristics.

2.2.1 Construction of traffic road network graph structure

Traffic prediction is a typical time series prediction problem [33], and its road network traffic flow data exhibits a high degree of periodicity, which provides a great deal of potential for traffic prediction. **Figure 6** shows the traffic data for the first week of December for individual toll stations on the Shaanxi Provincial Freeway, demonstrating a high degree of periodicity.

Given the first M flow observations, the flow data measured at the n sensor stations at time step H can be viewed as a matrix of size $M \times N$. The most likely flow measurements predicted at the next H time steps are:

$$F_{t+1}, \dots, F_{t+H} = argmaxlogP(F_{t+1}, \dots F_{t+H} | F_{t-M+1}, \dots, F_t) \tag{14}$$

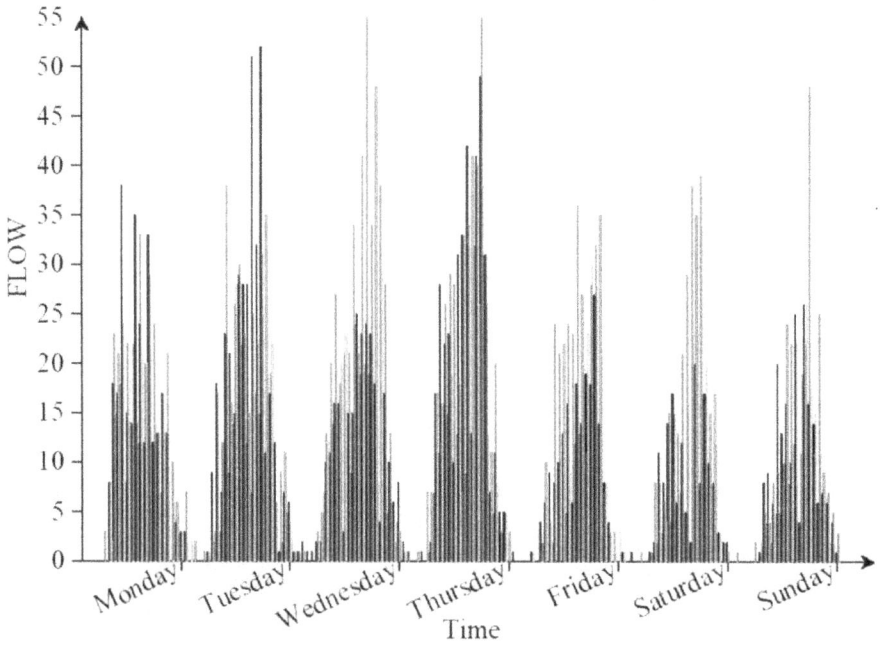

Figure 6.
Periodicity of traffic data.

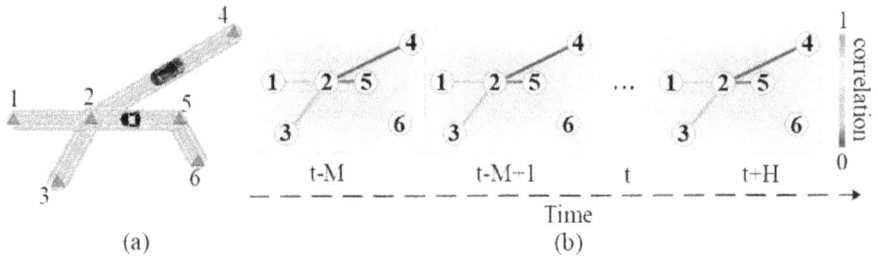

Figure 7.
Spatiotemporal correlations. (a) Stations in a road network. (b) Dynamic spatial correlations.

where $F_t \in R^n$ is a vector of observations for n road segments at time step t, where each element records the historical observations for a single road segment. For unordered road network traffic data, the observations F_t are not independent and can be viewed as graph signals defined on an undirected graph G with weight as shown in **Figure 7**, the graph is expressed in terms of an adjacency matrix $G_t = (V_t, E, W)$. F_t is a finite set of vertices corresponding to the observations of the n toll stations in the traffic. E is a set of edges representing the connections between stations, and $W \in R^{n \times n}$ represents the weighted adjacency matrix of G_t.

2.2.2 Traffic data acquisition and preprocessing

2.2.2.1 Traffic datasets

Traffic flow prediction by deep learning requires a large amount of data support, that is, real-world road traffic speed data. With the continuous improvement of traffic facilities, the amount of traffic data has also produced an explosive growth. Traffic flow prediction is precisely based on huge traffic data, so understanding the current common traffic data is the basis for achieving traffic flow prediction. The sources of traffic data mainly include road fixed-point detectors, vehicle GPS records, bus IC cards, license plate recognition, cell phone data, etc. We have made the common traffic data used for traffic flow prediction as **Table 1**.

2.2.2.2 Data preprocessing

Traffic flow data mainly detects parameters such as speed, flow rate, time, etc. The data collection process may result in detection equipment failure, instrument error, software failure, communication interference, environmental noise, etc., and even sudden road failure may have a great impact on the data, resulting in real-time data may be missing or abnormal, so the overall process of validity processing of this type of traffic data according to its type is shown in **Figure 8**.

• Abnormal data processing

The preprocessing methods of abnormal data can be divided into two categories: Data rejection. Data rejection can be used when there is less erroneous data in the traffic data. The rejection of individual erroneous data will not affect the integrity and trend of the data, but if the proportion of erroneous data is large, the rejection method cannot be adopted because too much rejection of erroneous data will

Data set classification	Data set name	Data fields	Sampling period
Expressway	PeMS	Timestamp, Station ID, Region. Highway ID, direction, trip	5 min
	METR-LA	Vehicle speed	5 min
	SEATTLE LOOP	Vehicle speed	5 min
	Madird Traces	Vehicle track	0.5 s
	Los-Loop	Vehicle speed	5 min
Cabs	NYC Taxi	Boarding and alighting times, location, distance traveled, fare, payment type	/
	TaxiBJ21	GPS data and weather data	30 min
	SH-Speed	Vehicle ID, location, operation status, speed	10 min
	CRAWDAD	Vehicle ID, time, coordinates	7 s
	SZ-taxi	Vehicle speed	15 min
	T-Drive	Vehicle track	/
Internet taxi	Didi-GAIA-Open-Data	Vehicle speed	/
Rail Transit	SHMetro HZMetro	People flow	15 min
City Road Network	VTC (vant-trace-creteil)	Time, lane, vehicle angle, speed, and vehicle ID	1 s
	0d_bologna Koln.Tr	Coordinates, speed, vehicle ID	1 s
	NYC-Bike	Vehicle ID, coordinates, time	/

Table 1.
Common data sets for traffic flow prediction models.

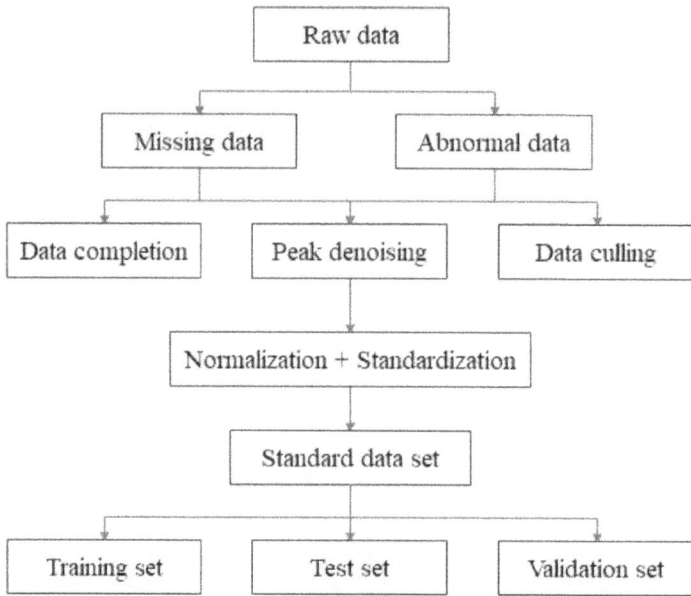

Figure 8.
Overall flow of data preprocessing.

destroy the integrity of the data and its trend. Peak denoising. Since traffic data is highly nonlinear and the traffic data at peak hours can be very significant, i.e., the noise oscillation region during peak hours, peak denoising is needed. Commonly used methods such as empirical mode decomposition (EMD), i.e., fluctuation decomposition in the local oscillation part of the trend change.

• Missing data processing [34]

Missing data is caused by hardware and software factors that do not detect data at the detection end or packet loss during data communication. In road traffic, this can be due to excessive vehicle density and inaccurate data collection by traffic flow detection instruments, data failures in transmission, and many other reasons for gaps in the collected data, such as missing data at a point in time, a certain period, or several periods of time. Typically, there are two classical missing patterns in time series data as shown in **Figure 9** below. **Figure 9(a)** indicates that the exported toll records have randomly lost observations at a single toll station, and the white circles indicate the missing values. **Figure 9(b)** indicates that there are several consecutive time points in the records of multiple toll stations with no observed values, which is

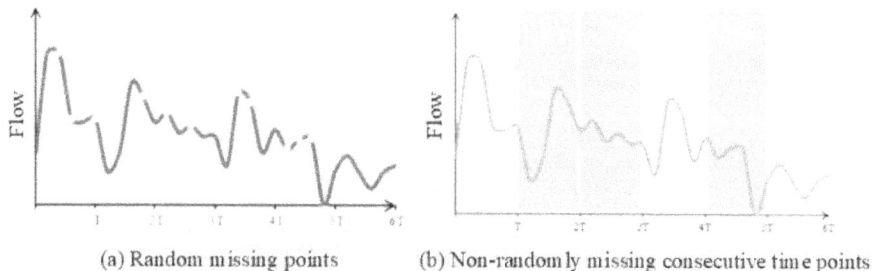

(a) Random missing points   (b) Non-randomly missing consecutive time points

Figure 9.
Example of missing pattern of spatiotemporal data (traffic data as an example).

a more common pattern of missing spatiotemporal traffic data. The green curve in the green panel represents the observed values and the gray curve represents the missing values. This situation requires correlating and processing the missing data, and then repairing the data using interpolation and smoothing algorithms, prior to dimensionlessizing the data using initialization operators to consider the fact that the units and orders of magnitude of the characteristic series of influencing factors are not uniform.

- Data normalization/normalization

Generally, the obtained traffic data are scattered, and the distribution characteristic curve presented by the data is fuzzy, and the distribution cannot be determined. Therefore, the data do not satisfy the normal distribution and need to be normalized to regularize the data and improve the comparability between the data to facilitate the subsequent model prediction. The data are z-core normalized to approximately satisfy the normal distribution, so that the weights are more evenly distributed in the subsequent model training, i.e.,

$$x' = \frac{x - \mu}{\sigma} \tag{15}$$

where μ, σ are the mean and standard deviation.

2.2.3 Classical graph convolution framework

To solve the problem of non-Euclidean structure of traffic network data, graph neural networks are often used to model spatial dependencies in traffic networks, and then convolution is used to fundamentally improve the efficiency of graph analysis and network construction from frequency and spatial domains, i.e., Graph Convolutional Network (GCN). Graph Convolution extends traditional convolution to graph-structured data, and powerful methods such as graph convolutional networks and their variants are widely used for these spatiotemporal network data prediction tasks with good performance. Most existing graph convolutional traffic flow forecasts are spatiotemporal in nature, since most traffic data sets have both spatial and temporal attributes. The development of traffic flow prediction models based on graph convolutional networks is presented as in **Figure 10**. In this paper, five of the most typical and most referenced models will be selected for illustration.

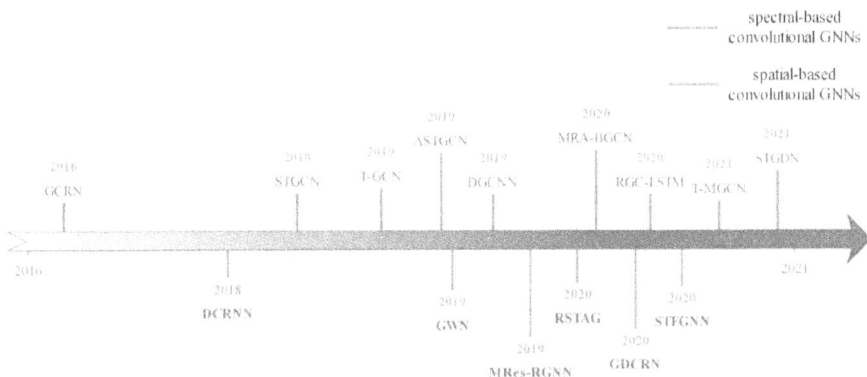

Figure 10.
Traffic flow prediction model based on graph convolution.

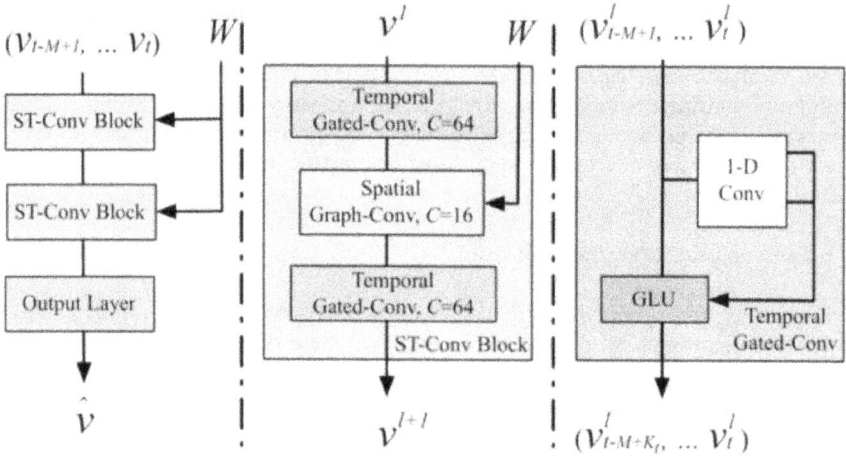

Figure 11.
Architecture of spatiotemporal graph convolutional networks.

2.2.3.1 STGCN predicts traffic flow

The STGCN model proposed by Yu et al. [19] (**Figure 11**) (2018AAAI) for the first time uses graph structures to model traffic networks while using graph convolution to model spatiotemporal sequences and uses pure convolutional structures to extract spatiotemporal features from the graph structures simultaneously.

STGCN is composed of several spatiotemporal convolutional blocks, each of which is formed as a "sandwich" structure with two gated sequential convolution layers and one spatial graph convolution layer in between. The framework STGCN consists of two spatiotemporal convolutional blocks (ST-Conv blocks) and a fully-connected output layer in the end. Each ST-Conv block contains two temporal gated convolution layers and one spatial graph convolution layer in the middle. The residual connection and bottleneck strategy are applied inside each block.

The model, although using convolution instead of LSTM-like patterns, does speed up training, but it also leads to missing historical data information and can only achieve short-term prediction, not long-term prediction, and graph convolution captures information between different nodes to model spatial models, which does not seem to make good use of the potential relationships between different regions.

2.2.3.2 DCRNN for predicting traffic flow

The DCRNN model proposed by Li et al. [22] (**Figure 12**) (2018ICLR) models spatial correlation as a diffusion process on directed graphs, thus modeling the transformation of traffic flow, and proposes diffusion convolution recurrent neural networks that can capture the spatial and temporal dependence between time series using a framework of seq2seq. To address these challenges, we propose to model the traffic flow as a diffusion process on a directed graph and introduce Diffusion Convolutional Recurrent Neural Network (DCRNN), a deep learning framework for traffic prediction that incorporates both spatial and temporal dependency in the traffic flow. Specifically, DCRNN captures the spatial dependency using bidirectional random walks on the graph and the temporal dependency using the encoder-decoder architecture with scheduled sampling.

2.2.3.3 *GMAN prediction of traffic flow*

The GMAN model (2020AAAI) proposed by Zheng et al. [23] (**Figure 13**) uses a spatiotemporal attention mechanism to model dynamic spatial relationships and nonlinear temporal relationships separately, while using a gating mechanism to

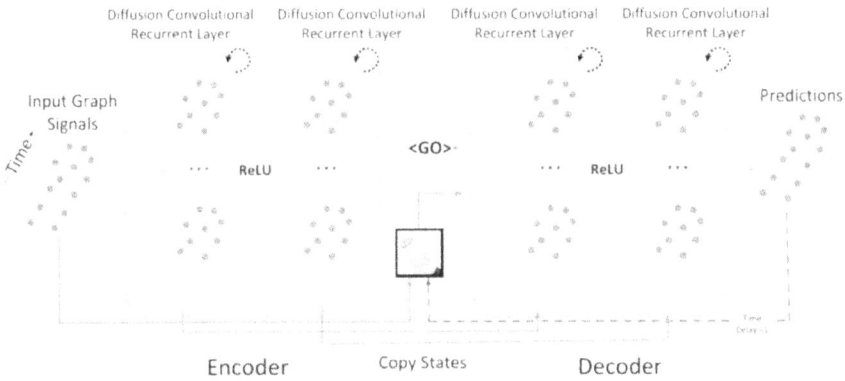

Figure 12.
System architecture for the diffusion convolutional recurrent neural network designed for spatiotemporal traffic prediction.

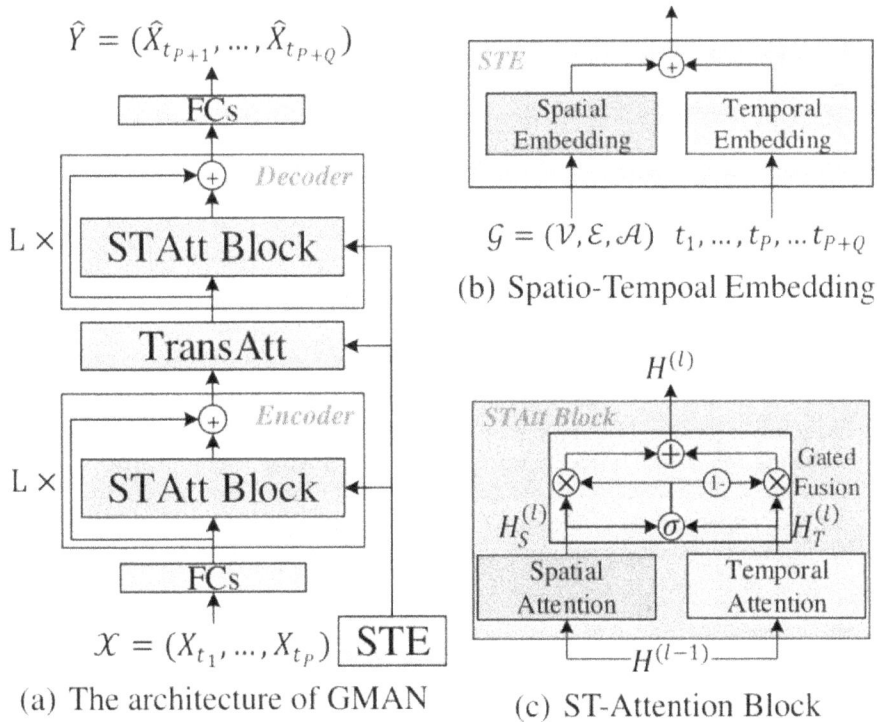

Figure 13.
The overall framework of GMAN model. (a) the framework of GMAN. (b) Spatiotemporal embedding. (c) the ST-attention block.

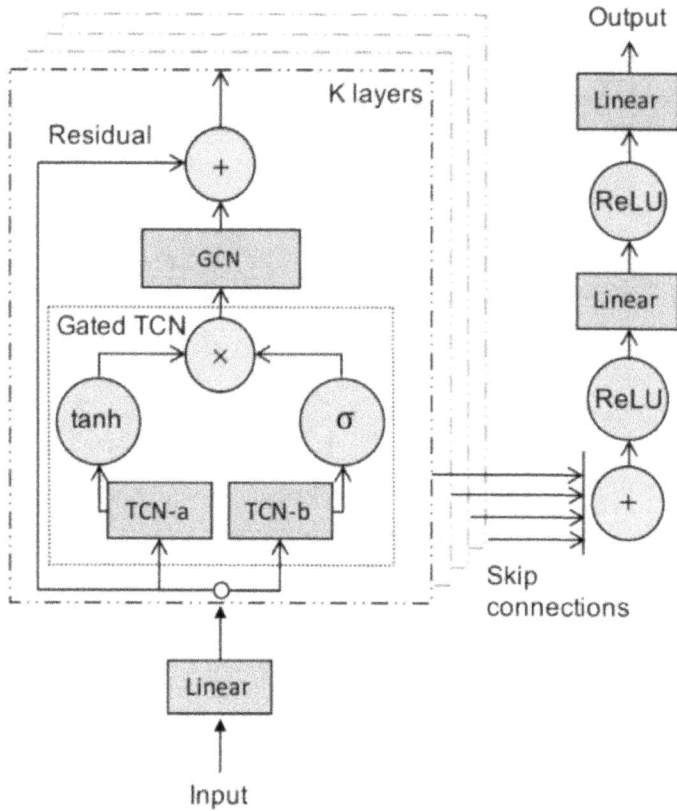

Figure 14.
The framework of graph WaveNet.

adaptively fuse the information extracted by the spatiotemporal attention mechanism.

Because the whole traffic is a network, the error of one node is amplified by other nodes, which affects the final prediction results. To solve the above problem, GMAN adopts an encoder-decoder architecture, where encoder is used to extract features and decoder to predict. A transformed attention layer is applied in between these two to transform the encoded traffic features to generate a sequential representation of future time steps as the input to the decoder. Here Encoder and Docoder are composed of ST-attention block. Then the authors use an STE block to combine the spatial and temporal information and then input into the ST-ATTENTION block to solve the problem of complex time–space correlation. Finally, the experimental results of the article on two real-world traffic prediction tasks (i.e., traffic volume prediction and traffic speed prediction) demonstrate the superiority of GMAN.

2.2.3.4 Graph WaveNet prediction of traffic flow

Wu et al. [35] (**Figure 14**) propose in this paper a novel graph neural network architecture, Graph WaveNet, for spatial–temporal graph modeling. The model uses the idea of diffusion convolution in extracting spatial features of road networks and adds a novel adaptive connection matrix to make up for the deficiency of fixed

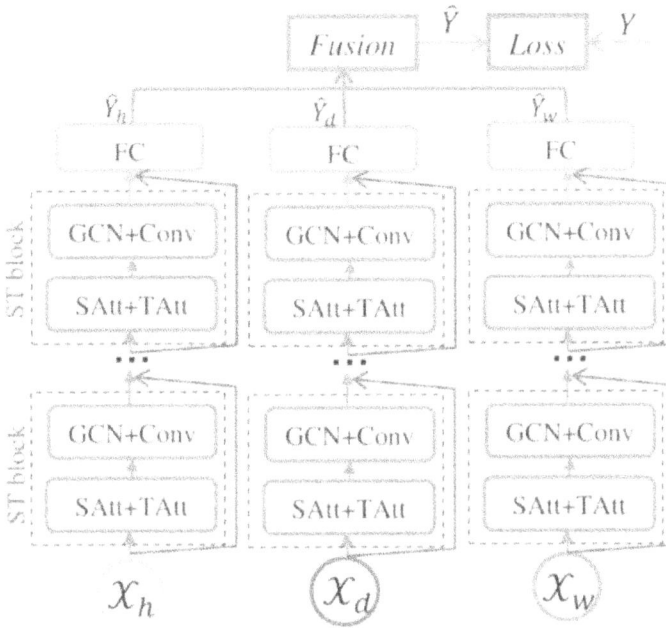

Figure 15.
The framework of ASTGCN. SAtt: Spatial attention.

topology in extracting spatial features and employs dilated causal convolution and gate mechanism on time series without the traditional RNNs cycle, which is validated by METR-LA and PEMS-BAY data sets, GWN in terms of training effect and time good results were achieved.

*2.2.3.5 **ASTGCN** prediction of traffic flow*

Guo et al. [24] (**Figure 15**) propose a novel attention-based spatial–temporal graph convolutional network (ASTGCN) model to solve traffic flow prediction problem. ASTGCN mainly consists of three independent components to respectively model three temporal properties of traffic flows, i.e., recent, daily-periodic, and weekly-periodic dependencies. More specifically, each component contains two major parts: 1) the spatial–temporal attention mechanism to effectively capture the dynamic spatial–temporal correlations in traffic data; 2) the spatial–temporal convolution, which simultaneously employs graph convolutions to capture the spatial patterns and common standard convolutions to describe the temporal features. The output of the three components is weighted fused to generate the final prediction results.

These five typical traffic flow prediction models above are compared in **Table 2**.

2.3 Application scenarios for traffic flow prediction

Many researchers have already applied the proposed traffic flow prediction models to various traffic scenarios and achieved excellent results. For example, by predicting the traffic flow of a roadway in advance, it can provide drivers with more advanced travel routes. In addition, it can also provide prerequisites for traffic light optimization, etc. In this paper, three scenarios are chosen to illustrate the

Models	Model's characteristics
STGCN (2018)	(1) Compared with traditional spatio-temporal models (RNN, LSTM) based on recurrent neural networks, the STGCN model combines graph convolution and gated time convolution, and for the first time uses pure convolutional layers to extract time and space information at the same time. (2) The STGCN model uses one-dimensional convolution to learn information in the time dimension and is not limited by the prediction data at the previous time point, so that the model can better capture the drastic changes in the data (such as traffic flow data during peak hours). (3) Due to the characteristics of the convolutional structure, STGCN model is parallelized at the input, with fewer parameters and faster training speed, allowing the model to process large-scale networks with higher efficiency.
DCRNN (2018)	(1) In view of the dynamic characteristics of traffic flow, DCRNN model introduces diffusion convolution when modeling spatial dependence and considers forward propagation and back propagation and is more suitable for traffic networks with a directed graph structure. (2) When the DCRNN model models the time dependence, the matrix multiplication in the GRU model is changed to diffusion convolution, and the diffusion convolution-gated recurrent unit (DCGRU) is obtained.
ASTGCN (2019)	(1) The ASTGCN model effectively learns the dynamic spatiotemporal correlation of traffic data through the spatiotemporal attention mechanism. (2) The ASTGCN model designed a multichannel network structure from multiple time periods, combining global time and space information to improve prediction accuracy
GWN (2019)	(1) In terms of spatial dependency acquisition, The GWN model constructs an adaptive dependency matrix that can retain the implicit spatial relationship of the road network. (2) In the acquisition of time dependence, The GWN model adopts stacked dilated 1D convolution and does not need to consider the problem of information disappearance too long ago and can extract longer time dependence than RNN-based cyclic convolutional networks.
GMAN (2020)	(1) The GMAN model refines the complexity of time and space. It is divided into dynamic spatial correlations and Non-linear temporal correlations, and the attention mechanism is introduced. (2) The GMAN model solves the cumulative error problem of stepwise prediction. A transformation attention layer is added between the encoder and the decoder, so that historical and predicted traffic characteristics can be converted.

Table 2.
Characteristics of typical models.

application of traffic flow prediction in the context of highways. These three scenarios are the work that has been done by our team so far, and the reason for choosing the highway is that the work in this area is more mature.

2.3.1 Scenario I. quantitative assessment on truck-related road risk for the safety control

Traffic conditions of truck flow is one of the critical factors influencing transportation safety and efficiency, which is directly related to traffic accidents, maintenance scheduling, traffic flow interruption, risk control, and management. The estimation of the truck flow of various types could be better to identify the irregular flow variation introduced by various trucks and quantitatively assessed the corresponding road risks.

Jin et al. [36] first improved on the gated recursive unit (GRU) based on a deep learning approach to estimate various types of truck traffic. Then a multiple logistic regression method was proposed to classify the road risk into three classes: safe, risky, and dangerous. According to the CSV trend, road risks are classified into

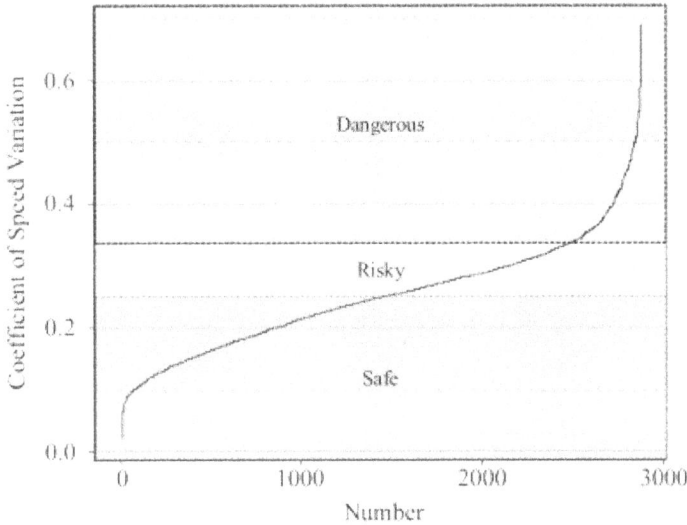

Figure 16.
Coefficient of speed variation (CSV) of passenger cars.

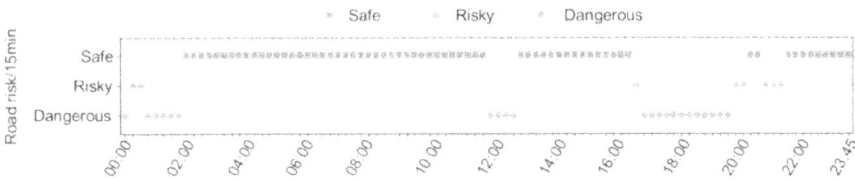

Figure 17.
Road risk assessed by predicted truck flow in April 12,018.

three categories as shown in **Figure 16**. Different risk classes can guide traffic control and management and broadcast traffic information to drivers to help them choose their travel routes.

Finally, the road risk calculated by the predicted truck traffic is shown in **Figure 17**, from which the road risk status can be obtained at every moment.

2.3.2 Scenario II. Improved manpower planning for highway toll gate

In China, the relatively heavy queues at freeway toll booths and service areas during peak hours, coupled with the saturation of manpower scheduled during off-peak hours, are undoubtedly a huge obstacle to efficient and cost-effective freeway operations. Therefore, it needs an intelligent manpower planning strategy to simultaneously ensure the efficiency of highway transportation management and road user satisfaction.

Jin et al. [37] addressed a high-precision prediction of vehicle flow based on historical multisource traffic data. Based on the prediction results, an improved manpower planning strategy is proposed to schedule the work accordingly. And the method was tested on a randomly selected toll station as an example, as **Figure 18** shows the daily traffic pattern of the highway Hechizhai toll station.

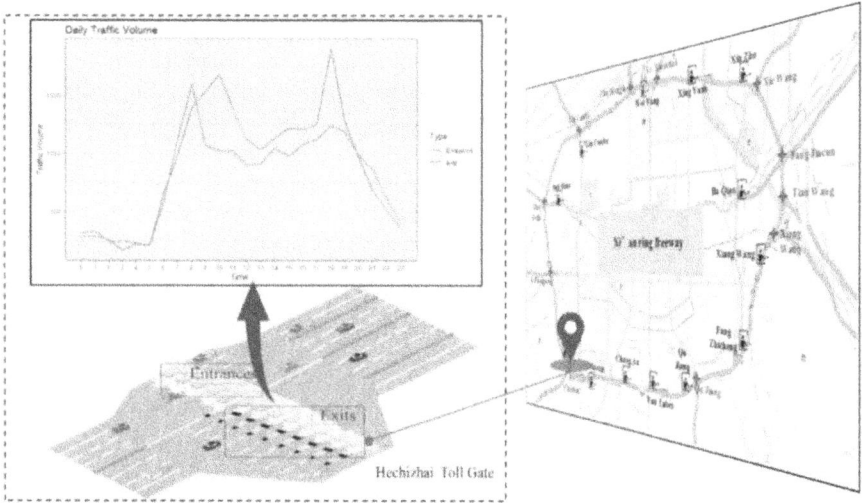

Figure 18.
Daily traffic pattern of the Hechizhai toll gate on the freeway.

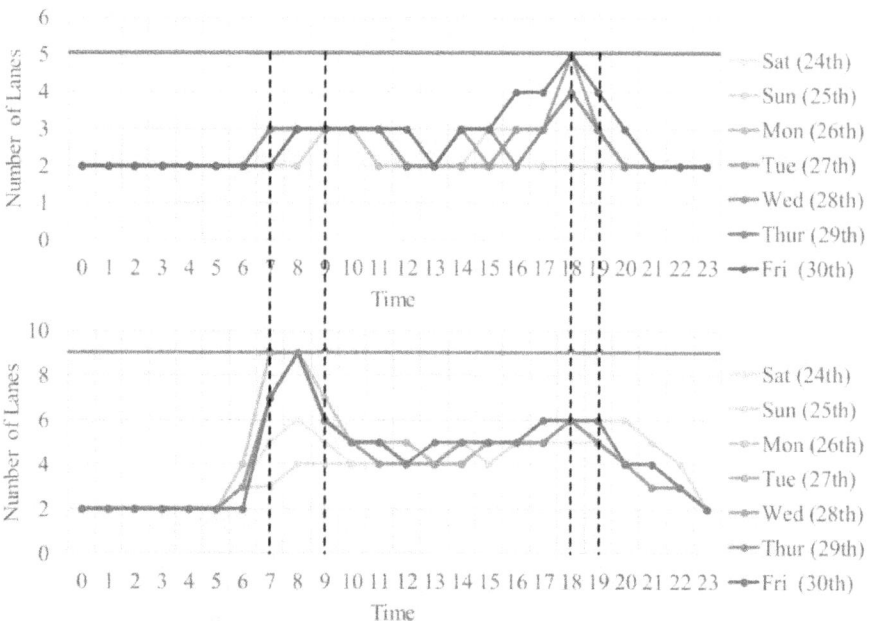

Figure 19.
Reversible toll lane configuration suggestion. The red lines refer to the number of toll lanes designed in each direction. The blue and brown gradient lines indicate the change in the number of exit and entrance lanes per hour, respectively. The black dash lines depict the reversible toll lane change period.

The results show from **Figure 19** that the upper part and the lower part show the lane opening at the entrance and exit of toll gates in one week, respectively. Two narrow black dotted lines indicate the morning and evening peak hours. During the morning peak, it is obvious that two entry lanes are not used while the number of toll lanes of the exit has reached its upper threshold. The opposite phenomenon is

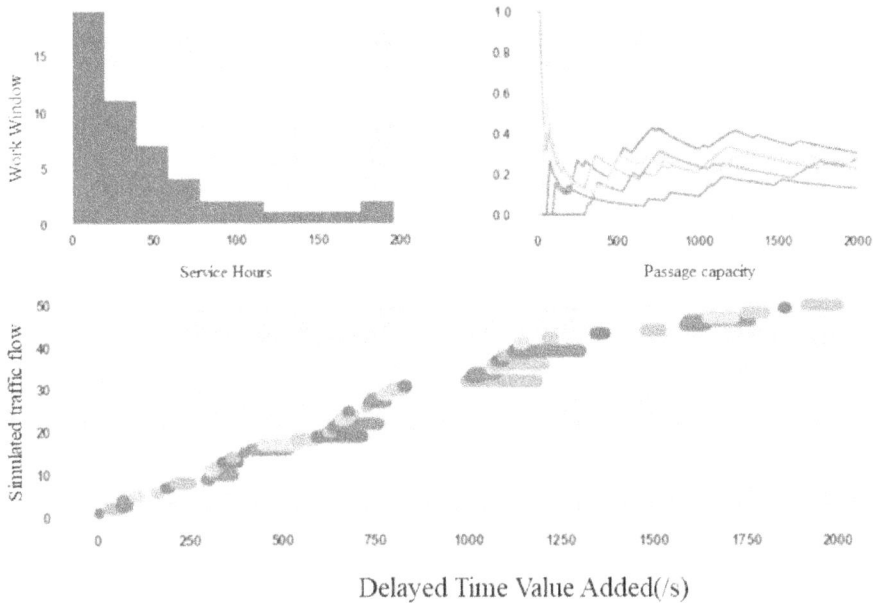

Figure 20.
Comparison of duplex lane simulation process.

seen in the evening peak. Therefore, we suggest that one or two of the entrance lanes can be set as a reversible lane so that the traffic pressure can be released in peak hours. Note that the usage condition of this suggestion is that the entrance and exit of the toll gate must be adjacent such as Hechizhai toll gate.

2.3.3 Scenario III. Capacity analysis of toll stations

Yuan et al. [38] used the results of traffic prediction to analyze the capacity of a toll station and used different queuing models to describe the capacity of typical lanes and compared the delay time and queue length of each model and obtained that the single-way model is more efficient in a typical system. The traffic index of the multiplex lane is also simulated, and the specific simulation process is shown in **Figure 20**, and the capacity of the multiplex lane is obtained to be larger than that of the typical MTC lane, which can relieve the traffic pressure during the peak hours.

3. Conclusions

Traffic is the main driving force of urban development, and real-time and accurate traffic flow prediction is the key to the application of intelligent transportation system. Graph convolutional neural network is an efficient model for processing graph data and has received a lot of attention from researchers in the past few years. This section attempts to summarize the recent graph convolutional neural network models and their applications to traffic flow prediction.

1. This section summarizes the GCN-based traffic flow prediction model. Starting from the basic definition of graph convolution, the basic principles of GCN are introduced with the focus on frequency-domain graph convolution

and space-domain graph convolution. Then, the representative models are clarified, and the structure and characteristics of different prediction models are further categorized and reviewed.

2. This section provides the traffic prediction problem with constructed traffic graph structure. Some public traffic data sets that are widely used in scientific research worldwide are introduced for traffic prediction experiments, including their data sources, data contents, and data acquisition addresses, and the whole data processing process is analyzed.

3. The application scenarios of traffic flow prediction are discussed. There are two applications are provided: 1) with prediction of the traffic flow of truck, the transportation safety and efficiency could further assess. 2) The work schedule arrangement could also improve based on the prediction of traffic flow to avoid manpower waste and allow more passing gate to open in the peak hours. Other than those applications, there still many aspects worth to explore.

Acknowledgements

Thanks to the following researchers for their great support in the writing of this book, especially Wenbang Hao, Erlong Tan, Wanrong Xu, Zhen Jia, Yiwen Gao, Yajie Zhang, etc., who have put a lot of energy into many formulas and illustrations.

This work was supported in part by the National Key R&D Program of China (2020YFB1600400), Key Research and Development Program of Shaanxi Province (No.2020GY-020), National Natural Science Foundation of China (Grant No. 51505037) and Supported by the Fundamental Research Funds for the Central Universities, CHD (Grant No. 300102320305).

Author details

Ping Wang*, Tongtong Shi, Rui He and Wubei Yuan
Chang'an University, Xian, China

*Address all correspondence to: pingwang@chd.edu.cn

IntechOpen

References

[1] Wei Chen H, Klaiber A. Does road expansion induce traffic? An evaluation of vehicle-kilometers traveled in China. Journal of Environmental Economics and Management. 2020;**104**:95-696. DOI: 10.1016/j.jeem.2020.102387

[2] Ye J, Zhao J, Ye K, Chengzhong X. How to build a graph-based deep learning architecture in traffic domain: A survey. IEEE Transactions on Intelligent Transportation Systems. 2020;**2**(6):1-20. DOI: 10.1109/TITS.2020.3043250

[3] Wang P, Hao W, Jin Y. Fine-grained traffic flow prediction of various vehicle types via fusion of multisource data and deep learning approaches. IEEE Transactions on Intelligent Transportation Systems. 2020;**5**:8-10. DOI: 10.1109/TITS.2020.2997412

[4] Tian Z, Jia L, Dong H, Zhang Z. Determination of key nodes in urban road traffic network. Shenyang, China: Proceeding of the 11th World Congress on Intelligent Control and Automation. 29 June-4 July 2014, IEEE; 2014. pp. 3396-3400. DOI: 10.1109/WCICA.2014.7053279

[5] Wang P, Xu W, Jin Y, Wang J, Li L. Forecasting traffic volume at a designated cross-section location on a freeway from large-regional toll collection data. IEEE Access. 2019;**7**:9057-9070. DOI: 10.1109/ACCESS.2018.2890725

[6] Jin Y, Xu W, Wang P. SAE Network: A Deep Learning Method for Traffic Flow Prediction. 2018 5th International Conference on Information, Cybernetics, and Computational Social Systems (ICCSS). 2018:241-246. DOI: 10.1109/ICCSS.2018.8572451

[7] Yu X, Sun L, Yang Y, Liu G. A short-term traffic flow prediction method based on spatial–temporal correlation using edge computing. Computers & Electrical Engineering. 2021;**93**:107219. DOI: 10.1016/j.compeleceng.2021.107219

[8] Miska MP. Microscopic Online Simulation for Real-Time Traffic Management. TRAIL Research School. 2007

[9] Ngoduy D, Wilson RE. Multianticipative nonlocal macroscopic traffic model. Computer-Aided Civil and Infrastructure Engineering. 2014;**29**(4):248-263

[10] Okutani I, Stephanedes YJ. Dynamic prediction of traffic volume through Kalman filtering theory. Transportation Research Part B: Methodological. 1984;**18**(1):1-11

[11] Kumar SV, Vanajakshi L. Short-term traffic flow prediction using seasonal ARIM A model with limited input data. European Transport Research Review. 2015;**7**(3):21

[12] Jeong YS, Byon YJ, Castro-Neto MM, Easa SM. Supervised weighting-online learning algorithm for short-term traffic flow prediction. IEEE Trans. on Intelligent Transportation Systems. 2013;**14**(4):1700-1707. DOI: 10.1109/TITS.2013.2267735

[13] Zhu S, Lin C, Chu Z. Bayesian network model for traffic flow estimation using prior link flows. Journal of Southeast University (English Edition). 2013;**29**(3):322-327. DOI: 10.3969/j.issn.1003-7985.2013.03.017

[14] Qi Y, Ishak S. A Hidden Markov Model for short term prediction of traffic conditions on freeways. Transportation Research Part C: Emerging Technologies. 2014;**43**:95-111. DOI: 10.1016/j.trc.2014.02.007

[15] Chen M, Yu X, Yang L. PCNN: Deep convolutional networks for short-term traffic congestion prediction. IEEE Transactions on Intelligent Transportation Systems. 2018;**19**(11): 3550-3559. DOI: 10.1109/TITS.2018. 2835523

[16] Lv Y, Duan Y, Kang W, Li Z. Traffic flow prediction with big data: A deep learning approach. IEEE Transactions on Intelligent Transportation Systems. 2015;**16**(2):865-873. DOI: 10.1109/ TITS.2014.2345663

[17] Hochreiter S, Schmidhuber J. Long short-term memory [J]. Neural Computation. 1997;**9**(8):1735-1780. DOI: 10.1162/neco.1997.9.8.1735

[18] Cho K, van Merrienboer B, Gulcehre C, Bahdanau D. Learning phrase representations using RNN encoder-decoder for statistical machine translation. Proceedings of the 2014 Conference on Empirical Methods in Natural Language Processing (EMNLP). 2014

[19] Yu B, Yin H, Zhu Z. Spatio-temporal graph convolutional networks: A deep learning framework for traffic forecasting. Proceedings of the Twenty-Seventh International Joint Conference on Artificial Intelligence. 2018. pp. 3634-3640

[20] Zhang J, Yu Z, Qi D. Deep spatio-temporal residual networks for citywide crowd flows prediction. In Proceedings of the Thirty-First AAAI Conference on Artificial Intelligence(AAAI'17). AAAI Press; 2017. pp. 1655–1661

[21] Pei H, Wei B, Chang KC-C, Lei Y, Yang B. Geom-GCN: Geometric Graph Convolutional Networks[J]. arXiv preprint arXiv:2002.05287, 2020

[22] Li Y, Yu R, Shahabi C, Liu Y. Diffusion convolutional recurrent neural network: data-driven traffic forecasting. arXiv preprint arXiv: 1707.01926, 2017

[23] Zheng C, Fan X, Cheng W, Qi J. GMAN: A graph multi-attention network for traffic prediction. Proceedings of the AAAI Conference on Artificial Intelligence. 2020;**34**(01): 1234-1241. DOI: 10.1609/aaai. v34i01.5477

[24] Guo S, Lin Y, Feng N, Song C, Wan H. Attention based spatial-temporal graph convolutional networks for traffic flow forecasting. Proceedings of the AAAI Conference on Artificial Intelligence. 2019;**33**:922-929. DOI: 10.1609/aaai.v33i01.3301922

[25] Kipf T, Welling M. Semi-supervised classification with graph convolutional networks. Published as a conference paper at ICLR. 2017. arXiv:1609.02907

[26] Liu Z, Zhou J. Introduction to Graph Neural Networks. Synthesis Lectures on Artificial Intelligence and Machine Learning. 2020;**14**(2):1-127

[27] Cheng Y, Liu Z, Cunchao T, Shi C, Sun M. Network Embedding: Theories, Methods, and Applications[J]. Synthesis Lectures on Artificial Intelligence and Machine Learning. 2021;**15**(2):1-242. DOI: 10.2200/S01063ED1V01Y20 2012AIM048

[28] Vassilis NI, Marques AG, Giannakis GB. Tensor Graph Convolutional Networks for Multi-Relational and Robust Learning. IEEE Transactions on Signal Processing. 2020; **68**:6535-6546. DOI: 10.1109/ TSP.2020.3028495

[29] Mestre Â. An Algebraic Representation of Graphs and Applications to Graph Enumeration. International Journal of Combinatorics. 2013;**2013**:347613, 14 pages. DOI: 10.1155/2013/347613

[30] Liao T, Wang W-Q, Huang B, Xu J. Learning laplacian matrix for smooth signals on graph. IEEE International Conference on Signal, Information and Data Processing (ICSIDP). 2019;**2019**: 1-5. DOI: 10.1109/ICSIDP47821.2019. 9173468

[31] Shi J, Cheung M, Du J, Moura JMF. Classification with vertex-based graph convolutional neural networks. 52nd Asilomar Conference on Signals, Systems, and Computers. 2018;**2018**: 752-756. DOI: 10.1109/ACSSC. 2018.8645378

[32] Tang S, Li B, Yu H. ChebNet: Efficient and Stable Constructions of Deep Neural Networks with Rectified Power Units using Chebyshev Approximations. arXiv. 2019

[33] Xiao J, Xie Y, Wen Y. The short-time traffic flow prediction at ramp junction based on wavelet neural network. IEEE 5th Advanced Information Technology, Electronic and Automation Control Conference (IAEAC). 2021;**5**:664-667. DOI: 10.1109/IAEAC50856.2021.9390960

[34] Yang H, Pan Z, Tao Q. Online learning for time series prediction of AR model with missing data. Neural Processing Letters. 2019;**50**(3): 2247-2263. DOI: 10.1007/s11063-019-10007-x

[35] Wu Z, Pan S, Long G. Graph wave net for deep spatial-temporal graph modeling. Twenty-eighth International Joint Conference on Artificial Intelligence IJCAI. arXiv preprint arXiv: 1906.00121, 2019

[36] Jin Y, Jia Z, Wang P, Sun Z, Wen K, Wang J. Quantitative Assessment on Truck-Related Road Risk for the Safety Control via Truck Flow Estimation of Various Types. IEEE Access. 2019;7: 88799-88810. DOI: 10.1109/ ACCESS.2019.2924699

[37] Jin Y, Gao Y, Wang P, Wang J, Wang L. Improved Manpower Planning Based on Traffic Flow Forecast Using a Historical Queuing Model. IEEE Access. 2019;7:125101-125112. DOI: 10.1109/ ACCESS.2019.2933319

[38] Yuan W, Wang P, Yang J, Meng Y. An alternative reliability method to evaluate the regional traffic congestion from GPS data obtained from floating cars. IET Smart Cities. 2021;**3**(2):79-90. DOI: 10.1049/smc2.12001

Section 4

Intelligent Device

Low-Cost Simple Compact and Portable Ground-Penetrating Radar Prototype for Detecting Improvised Explosion Devices

Krishnendu Raha and Kamla Prasan Ray

Abstract

This chapter presents the design and fabrication of a low-cost continuous-wave ground-penetrating radar for detecting improvised explosive devices buried in the soil for use of security forces. It is low cost and simple because it uses a single frequency (920 MHz) and is designed only for the detection of the buried target. The work presented includes designing of transmitter system module, receiver system module, antennas, power module, graphical user interface module, and making a prototype that is compact and portable. The chapter explains the concept and illustrates a method to enhance isolation between antennas, which is a very important parameter for the effective functioning of ground-penetrating radar. The presented method of enhancing isolation by cavity-backing a rectangular microstrip antenna and keeping them separated at an optimum gap yielded high isolation of 52.6 dB. The prototype radar, using the enhanced isolation antennas, demonstrates the capability to detect up to the depth of 65 cm for a circular steel target of radius 12.5 cm buried in loose semi-dry pebbled soil. The prototype radar is sensitive enough to detect a plastic box, a small bunch of wire, a book (paper) buried in soil and a wooden slab and a steel scale buried in a sandpit.

Keywords: ground-penetrating radar, improvised explosion devices, cavity-backed antenna, high-isolation antenna, low-cost prototype

1. Introduction

Ground-penetrating radar (GPR) uses the principle of scattering electromagnetic waves for its operation. A target buried in soil will have different dielectric properties as compared to soil. This change in dielectric constant will cause a change in amplitude and phase of the signal reflected from the soil with the target as compared to the signal reflected only by the soil. Thus, identifying the change in phase and amplitude of the reflected signal will assist in detecting a target buried in soil [1]. This simple theory may be utilised to design and fabricate a GPR prototype for detecting improvised explosive devices (IEDs) buried in the soil.

Studies related to GPR are being conducted in all advanced countries for almost about three decades. These studies resulted in the development of methodologies that can be employed to develop rugged, portable, and multipurpose GPR.

However, all these studies and methodologies aim at producing GPR, which can detect targets, ascertain the depth of the target and provide its pseudo-image [2–8]. The final GPR product become complex and cost-prohibitive incorporating all these functions concurrently.

For small detachments of security forces deployed in remote operational areas, the primary requirement of GPR is only to detect the target buried in soil, especially the IEDs. The other functionalities, such as determining the depth of the target and providing the pseudo-image of the target improve the detection value of the GPR, however, they make the system complex and cost-prohibitive and thus making it difficult for all the small detachments to procure them. The requirement is felt to design a low-cost simple GPR system that can only detect IEDs buried in the soil. This chapter intends to present a design and develop a prototype of a low-cost simple GPR for detecting IEDs buried in the soil. To make the product simple and low cost, only capable of detecting buried targets, the proposed system operates on a single frequency as opposed to wideband frequencies used in other commercial products.

The work presented in this chapter includes designing a continuous-wave transmitter and receiver module at a centre frequency of 920 MHz, designing a microcontroller-based module for detection of phase and amplitude variation, designing appropriate transmitter and receiving antennas with enhanced isolation between them, designing a user-friendly frontend and display and make the system online. The final objective of the work is to produce a portable prototype of GPR, which can detect IEDs buried in soil, for use of security personnel. The product design has been validated to detect both metals and non-metals buried objects in different kinds of soils and sand. The product is sensitive enough to detect a small bunch of wire and the maximum depth of detection achieved is 65 cm in loose semi-dry soil for a circular steel target of a radius of 12.5 cm. These experiments demonstrate that the prototype fabricated is capable of detecting IEDs buried in the soil.

The focus of the work is to design a low-cost GPR and use it only for the detection of the buried target. Thus, instead of operating using a wideband of frequencies [2–8] which provide other functionalities, such as providing exact depth and pseudo-image of the target, only a single-frequency operation is chosen to make the system simple and low cost capable of only detecting the target. A major problem in detecting a small target by the GPR is the masking of the low-power target reflected signal by relatively high mutual coupling between the antennas. To solve this issue, software pre-processing techniques, such as Background subtraction algorithm [9, 10], Displacement-based technique [11, 12], blind sources separation (BSS) techniques [13, 14], and hardware-based pre-processing techniques, such as filtering [15, 16], antenna polarisation technique [17, 18] and time gating techniques [19, 20], have been reported. These techniques make the system complex and cost-prohibitive too. This chapter explains in detail a simple low-cost but highly effective technique of enhancing isolation between transmitting and receiving antennas to resolve the issue of mutual coupling between antennas affecting the detection capability of the GPR. It is demonstrated that by introducing a cavity-backing on a rectangular microstrip antenna a destructive interface between direct and scattered radiation from the cavity rim yields maximum isolation for an optimised combination of cavity height and separation between two antennas. The cavity-backing yields maximum isolation of 71.4 dB and a minimum of 49.1 dB within a narrow BW of 5% for the centre frequency of operation at 920 MHz. The use of this proposed antenna with enhanced isolation makes the GPR highly sensitive and effective. The cavity-backing technique has demonstrated enhancing isolation in wideband operation also in which a double cavity-backing yielded uniform high isolation of more than 40 dB for a BW of 64% for the centre frequency of operation of 2.6 GHz.

Following sections of the chapter cover system implementation, the concept of enhancing isolation between antennas and implementation of the same, prototype fabrication, and results of various experiments and discussions.

2. System implementation

The design and implementation process of a homodyne continuous-wave (CW) GPR to be used in detecting IEDs is discussed in this section. The frequency of 920 MHz has been used as the operating frequency for the product designed because, at this frequency, the depth of detection can be nearly 1 m with good resolution [3]. The work intends to satisfy the optimum hardware requirement of GPR at the aforementioned frequency and hence concentrates on the design and integration of oscillator, filter, power divider, in-phase, and quadrature-phase (IQ) demodulator, antenna, etc. A block diagram depicting the various components of the designed system is shown in **Figure 1**. The system consists of a transmitter sub-section, a receiver subsection, antennas, a data acquisition system, and an online display system which has been elaborated on in the following sub-sections.

A single PCB has been designed for the transmitter, receiver, and microcontroller to make the system compact. The PCB is fabricated on a 1.6 mm FR-4 substrate with ε_r = 4.4 and tanδ = 0.02. The PCB designed along with all the components soldered is depicted in **Figure 2(a)**. **Figure 2(b)** depicts Arduino Uno (microcontroller) and the 6 V DC, 48 AH battery connected to the PCB. Designing a transmitter, a receiver, and a data acquisition sub-system are discussed in the following subsections.

2.1 Transmitter sub-system

The transmitter comprises a voltage-controlled oscillator (VCO), which is designed to generate 920 MHz, as shown in **Figure 2(a)**. The VCO unit has been designed using the BFP520 transistor in the Colpitts oscillator configuration [21]. BBY52 varactor diodes have been used for varying frequencies. A resistive power

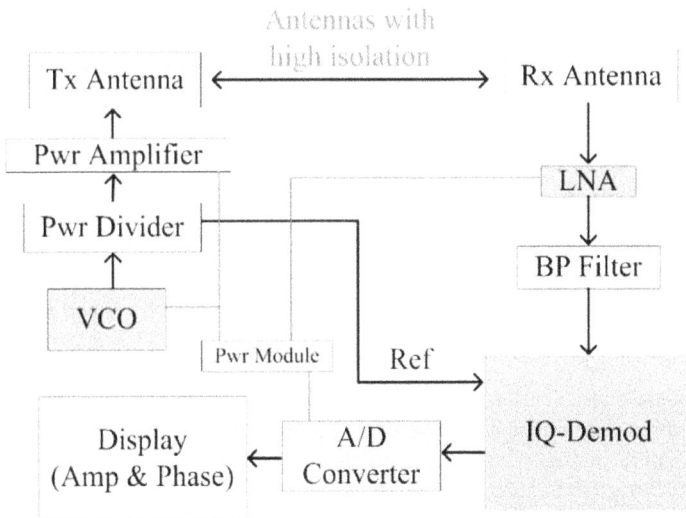

Figure 1.
Block diagram of the proposed low-cost GPR.

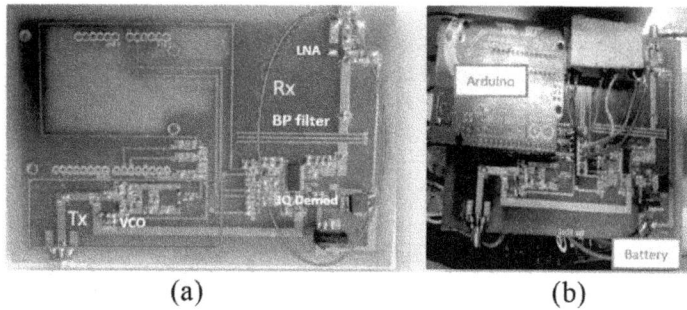

Figure 2.
PCB design (a) All components soldered (b) Ardiono and battery connected.

divider designed sends a part of this signal generated to the I-Q demodulator, which is used as a reference signal while demodulating.

2.2 Receiver sub-system

It receives the reflected signal from the ground and passes it through the low noise amplifier (LNA) for amplification. SGL 0622 LNA [22] has been used which provides a noise figure (NF) of less than 1.5 dB at 920 MHz and offers a gain of 30 dB. Inductive biasing is used to reduce noise. Following the LNA, a microstrip based three pole bandpass filter [23] is placed which is designed at the centre frequency of operation. Following this bandpass filter, a demodulator acting as phase and amplitude detector is placed. AD8347 [24] is used as a direct quadrature demodulator. It receives a reference signal from the transmitter end and provides amplitude ratio and phase difference between the transmitted and received signal. The amplitude and phase information are digitised, and the information obtained is displayed on a laptop.

2.3 Antenna design

Two identical transmitting and receiving rectangular microstrip antennas (RMSAs) resonating at 920 MHz have been designed using formulas given in Ref. [25]. The design of antennas augmented with the cavity backing to enhance the isolation between transmitting and receiving antennas is given in detail in Ref. [26]. It is of paramount importance to have enhanced isolation between the transmitting and receiving antenna so that a weak target reflected signal is not masked by the comparative high mutual coupling between the antennas.

The concept used to enhance isolation between the antennas is that when a cavity is introduced in co-located antennas, it makes two RF coupling paths between the two co-located antennas; one is direct and the other is via the cavity wall, as depicted in **Figure 3(a)**. With an optimised height of the cavity rim at a given separation between antennas, comparable electric fields from these two paths can be made out of phase (180°), cancelling each other, which can lead to maximising isolation between them. However, this technique will work only for narrowband operation because coupling path length is dependent on operating wavelength. For wideband operation, a multi cavity-backed structure is required where each cavity provides optimum path length for a particular narrow band of frequencies.

Extensive simulations have been done using Microwave CST software to find this combination of optimum cavity height (h) and separation (x) between the antennas which yield maximum isolation. **Figure 3(b)** depicts a schematic diagram

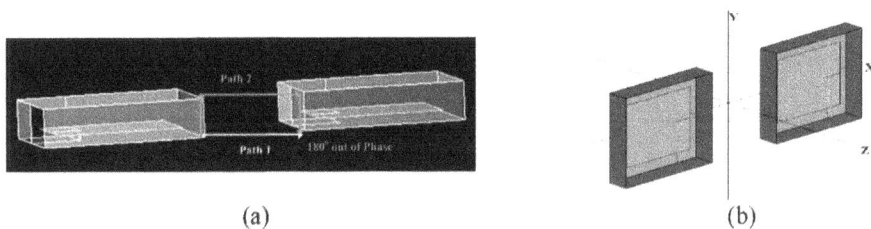

(a)   (b)

Figure 3.
(a) Concept used to enhance isolation between antennas (b) Simulated structure of two cavity-backed antennas for measuring isolation (S_{21}) between them.

Ser. no	x (mm)	$|S_{21}|$ (dB)	Ser no	x (mm)	$|S_{21}|$ (dB)
i.	30	35	vi.	120 (0.36 λ)	54.6
ii.	60	41.1	vii.	130	51.26
iii.	90	52	viii.	150	50
iv.	110	52.5	ix.	210	47.3
v	116	52.5	x.	240	46.8

Table 1.
Isolation (S_{21}) obtained by keeping the cavity height (h) to 40 mm and varying the separation (x) between the co-located antennas at 920 MHz [26].

Ser. no	h (mm)	$|S_{21}|$ (dB)	Ser No	h (mm)	$|S_{21}|$ (dB)
i.	20	33.7	vi.	42	47.4
ii.	26	36.5	vii.	43	45.3
iii.	30	39.9	viii.	50	37
iv.	40 (0.12 λ)	54.6	ix.	60	32.2
v	41	−50	x.	80	28.04

Table 2.
Isolation (S_{21}) obtained by keeping the separation (x) between the antennas fixed to 120 mm and varying the cavity wall height (h) at 920 MHz [26].

of two cavity-backed microstrip antennas with a separating distance of x for simulating isolation. **Tables 1** and **2** give the results of these simulations. The work has been elaborated in Ref. [26].

From the tables, it is noted that maximum isolation is obtained only at a particular combination of the height of the cavity wall and separation between the antennas. For the cavity height of h = 40 mm and separation of x = 120 mm, the isolation obtained is 54.6 dB. It is noted that the isolation is not enhanced either by increasing or decreasing the separation (of 120 mm) nor by changing the optimum cavity height (40 mm) for the 120 mm gap. It is, therefore, inferred that at the operating wavelength, destructive interference between direct and scattered radiation from the cavity rim yields maximum isolation for this combination of cavity height and separation between two antennas.

The top view and cross-sectional side view of the proposed antenna geometry are given in **Figure 4(a)**. The CST microwave software has been used to optimise the parameters. Optimised ground plane (GL × GW) of the antenna is 19.5 × 19.5 cm² and the radiating patch (L × W) is 13.5 × 13.5 cm² with the coaxial feed located

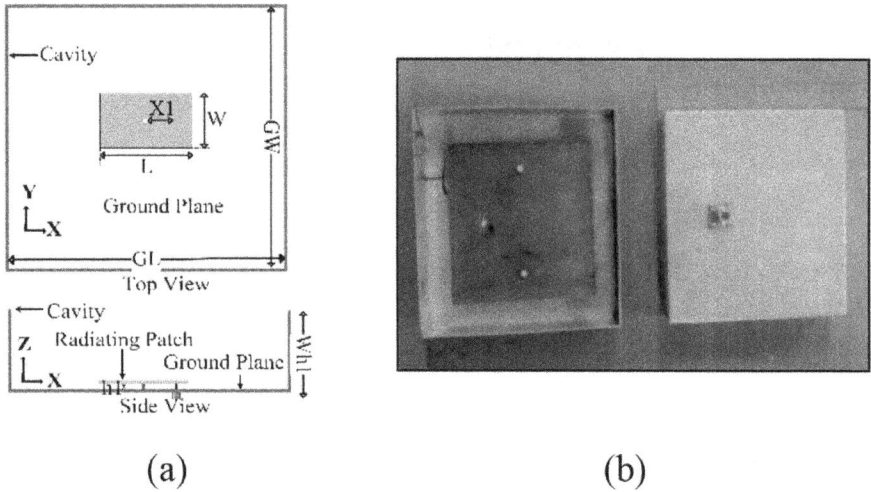

(a)   (b)

Figure 4.
(a) Top and cross-sectional side view of the proposed antenna configuration (b) photograph of fabricated cavity-backed antennas [26].

4.1 cm from the centre. A radiating patch is suspended in the air at a height (h1) of 1.4 cm from the ground plane. A cavity wall surrounds the patch antennas. The height (Wh1) of the cavity wall is taken to be 4 cm as obtained from **Tables 1** and **2**. Using these optimised design parameters, prototype transmitter and receiver cavity-backed RMSAs, as shown in **Figure 4(b)**, are fabricated. The radiating patch is made of copper plate suspended in the air with two Teflon supports at the centre line, and the ground plane and the cavity backing are made of aluminium. A comparison of measured and simulated isolation between the rectangular microstrip antenna without cavity-backing and with cavity-backing is given in **Figure 5**.

It is depicted in **Figure 5** that the introduction of cavity-backing improves the isolation by 25 dB at 920 MHz. Measured isolation with and without cavity wall for **0.36 λ_0** (12 cm) separation is 52.6 dB and 27.5 dB, respectively. Within 5% designed BW, cavity backing of the RMSA yields maximum isolation of 71.4 dB and

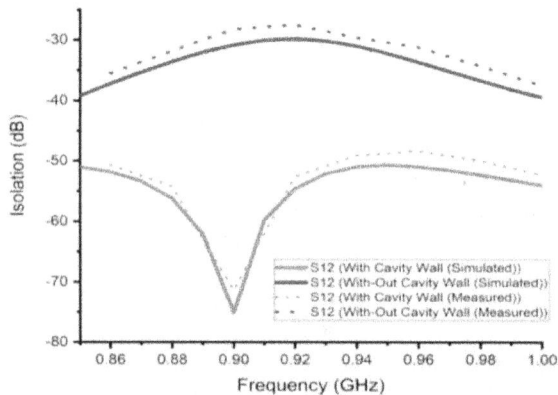

Figure 5.
Measured and simulated isolation at 0.36 λ_0 separation of cavity-backed RMSAs for a cavity wall height of 0.12 λ_0 [26].

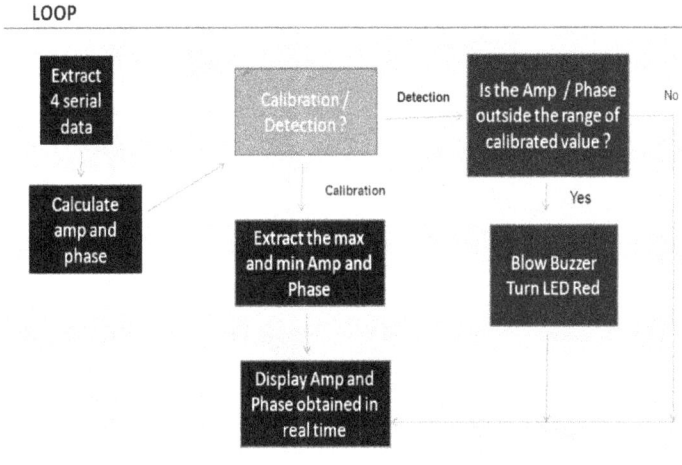

Figure 6.
Flow chart for online-display and calibration of the proposed low-cost GPR.

a minimum of 49.1 dB, whereas without cavity backing RMSA provided maximum isolation of 29.5 dB and a minimum of 27.5 dB only.

2.4 Data acquisition and online display system

The I and Q information of the reflected signal are passed through a low-pass filter with a cut-off at 20 Hz before feeding them to the A/D converter. In this work, an open-source Arduino Uno module [27] has been used as an A/D converter and microcontroller. For displaying amplitude and phase information of the target detected in real time a graphical programming environment (LabVIEW) has been used. LabVIEW design software has been integrated with Arduino [28] and a user-friendly graphical user interface (GUI) has been designed as depicted in the result section, to demonstrate the results. GUI design can calibrate the system according to soil conditions, fine-tune the frequency, and control the overall gain of the receiver. A basic flow chart for displaying amplitude and phase information in real time and designing the GUI with detection and calibration features is shown in **Figure 6**.

3. Fabrication of the GPR prototype

After fabrication and assembly of the single PCB and the integration of the power supply module, the final GPR system was integrated in such a way that there in minimum electromagnetic interference (EMI) amongst various subsystems. The unit was made compact and portable. In the final product, antennas are placed close to the ground at the base of the product. The antennas are grooved inside solid foam as shown in **Figure 7(a)**. The separation between two antennas is fixed at 09 cm to maximise isolation between them. As obtained from **Table 1**, 09 cm separating gap yields 52 dB isolation (sim) which is comparable to the maximum isolation yield of 54 dB (sim) for a separating gap of 12 cm. Thus, to keep the overall system compact, the separating gap between the antennas is kept at 09 cm. The solid foam is sprayed with zinc oxide paint so that it acts as an EM radiation absorber and increases the isolation between the antennas. The PCB and the battery are placed at the back (the front surface is the surface facing the ground where

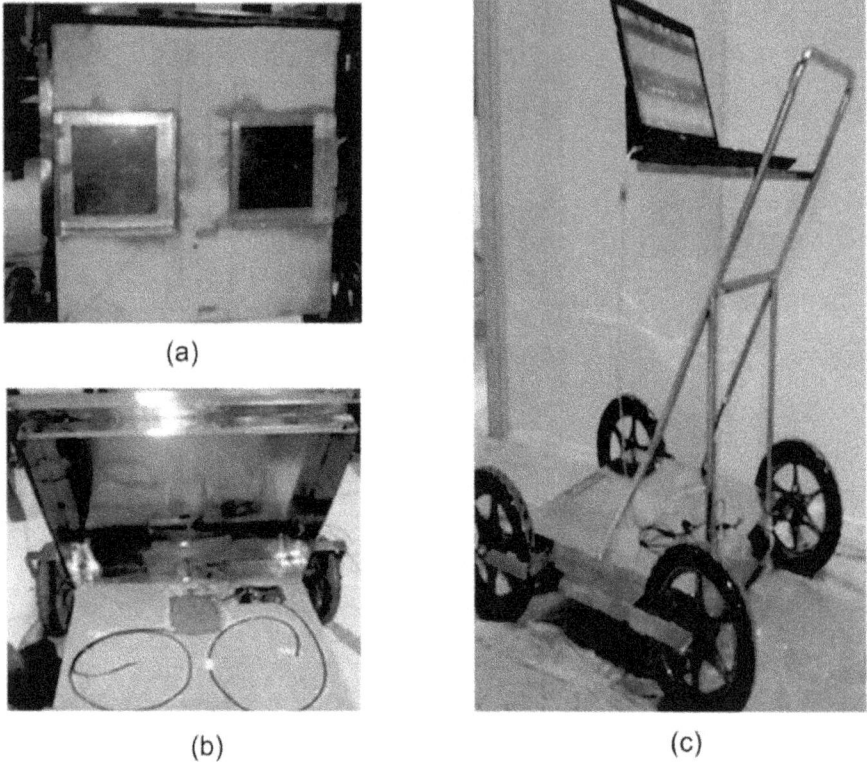

(a)

(b)   (c)

Figure 7.
Prototype of the fabricated GPR (a) front view of the base (b) back view of the base (c) final product with antennas, PCB, and power module at the base close to the ground rested on wheels and the display unit close to the user.

antennas are placed) ensuring that the back lobe of the antennas does not interfere with them. The placement of the PCB and battery is depicted in **Figure 7(b)**. The final product design is shown in **Figure 7(c)**.

4. Experiment with targets buried in soil and sand

To test the detection capability of the prototype GPR, various targets were buried in the ground. The first experiment was carried out to establish the maximum depth of detection of the GPR outside the laboratory in semi-dry soil. IEDs can be made of metals and non-metals and may be buried in different kinds of soils. Thus, experiments were conducted to detect a plastic box, a small bunch of wire, and a book (paper) buried in soil and a wooden slab and a steel scale buried in a sandpit.

4.1 Experiment to determine maximum depth of detection of the GPR

To measure the maximum depth of detection of the product, an experiment as depicted in **Figure 8(a)**, has been carried out. A circular steel target of a radius of 12.5 cm was buried in loose semi-dry soil with lots of small pebbles in it. The figure shows the target exposed but it has been buried in soil during the experiment such that the depth of soil over the target is about 65 cm. Next, the GPR prototype has been moved over the soil heap. **Figure 8(b)** depicts the detected signal. With

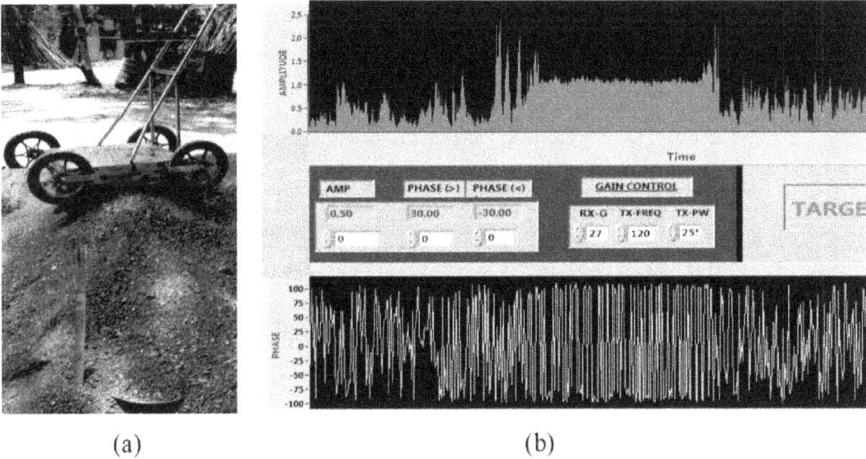

(a)   (b)

Figure 8.
Determining the maximum depth of detection of the GPR (a) experimental setup (b) GUI screen-shot for detection of a target at depth of 65 cm in the soil.

no target present, the detected amplitude level varies from 0.3 V to 0.8 V. This variation is because of the presence of many small pebbles in the soil. When the unit moves and reaches the location below which the target is buried, the received amplitude becomes stable at the level of 1 V. After this, as the prototype is moved away from the target the amplitude level again starts varying. Also, it is noted that because the target is buried so deep and the target size is not so big, the amplitude level detected as compared to the reference level (i.e., when no target is present) is not much different (only it is more stable) and no phase information about the target is obtained. It is inferred that the GPR cannot detect a target smaller than the present one beyond the depth of 65 cm in this type of soil.

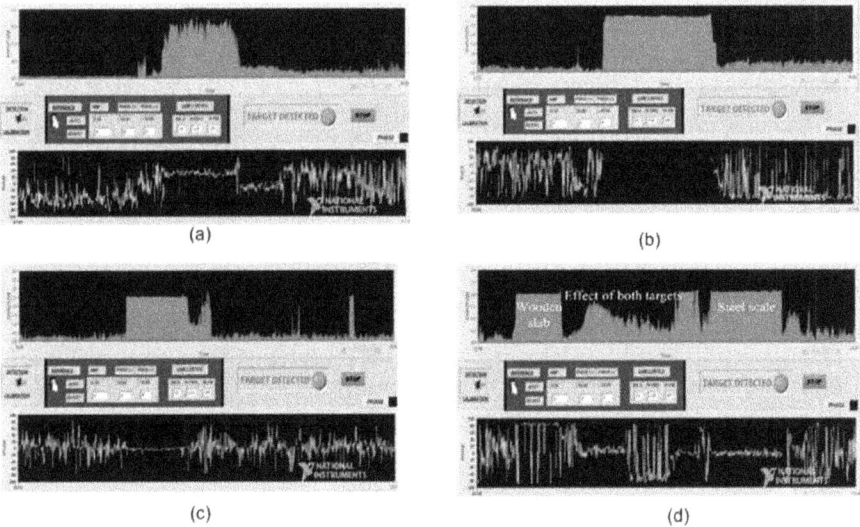

(a)   (b)

(c)   (d)

Figure 9.
GUI screen-shots of the received amplitude and phase response of targets buried in soil and sand (a) plastic box buried 20 cm in soil (b) bunch of wire buried 20 cm in soil (c) book (paper) buried 20 cm in soil (d) A wooden target (buried at 15 cm) and a steel target (buried at 30 cm) in a sandpit.

Ser. no.	Target type	Face area of target (cm^2)	Land type	Detection depth (cm)
(i)	Circular steel plate	490	Pebbled semi-dry soil	65
(ii)	Plastic box	150	Semi-dry soil	20
(ii)	Bunch of wire	110	Semi-dry soil	20
(iv)	Book (paper)	375	Semi-dry soil	20
(v)	Wooden slab	100	Sandpit	15
(vi)	Steel scale	225	Sandpit	30

Table 3.
The detection capability of the prototype GPR at 920 MHz.

4.2 Detecting targets buried in soil and sand

After determining the maximum depth of detection of the GPR, various experiments have been conducted to ascertain its capability to detect IEDs buried in the ground. The blast effect of IEDs depends on their size i.e. explosive content, depth at which it is hidden, and type of medium in which it is kept. A particular target kept at the same depth will have more blast effect when kept inside sand than in soil. For the same kind of medium, a smaller target hidden at lesser depth may have the same blast effect as compared to a relatively bigger target kept at greater depth. Keeping these blast effects of IEDs into consideration, to make the experiments assess the detection capability of the GPR prototype in practical scenarios, the following experiments have been conducted for metal and non-metal targets.

A small plastic box of size 15 × 10 × 3 cm^3, a bunch of wire, and a book (paper) of size 25 × 15 × 3 cm^3 have been used as targets and placed at a depth of 20 cm inside the soil. **Figure 9(a)** depicts the result obtained using GPR for the plastic box target. The high-reflected power obtained here is on account of the difference between the dielectric constant of air (trapped in the plastic box) and the dielectric constant of the soil. **Figure 9(b)** depicts the result for detecting a bunch of wires. In this case, the reflected power is not stable because it is a bunch of wires, that is, it consists of many small plastic-coated copper wires with soil in between and is not a monolithic big target. **Figure 9(c)** depicts that the product can also detect paper buried in the soil. Next, the medium in which the targets are buried is changed. A dry wooden slab (10 × 10 cm) and a steel plate (15 × 15 cm) have been buried at a depth of 15 cm and 30 cm, respectively in a sandpit with a 50 cm horizontal separation between them. The result obtained is shown in **Figure 9(d)**, which depicts that the product can detect both the targets buried in a sandpit.

4.3 Demonstrated detection capability

The detection capability demonstrated by the prototype GPR is given in **Table 3**. The table summarises the results of experiments conducted with the GPR operating at 920 MHz on the detection of various targets.

5. Discussion

The prototype single-frequency CW GPR successfully detected metal targets as small as a bunch of wire buried at 20 cm in soil and non-metal, such as wood, paper, and plastic, buried in the soil. For a metallic circular plate of a radius of 12.5 cm

Refs.	Centre frequency (GHz)	Isolation (dB)	Isolation method	Comments
[29]	10.0	40	RF absorber	Cost prohibitive
[30]	2.5	50	Spatial notch	complex design
[31]	5.2	44	Resonator between antennas	Narrowband & complex
[32]	2.6	50	Metamaterial cavity	Inherent narrow band
[33]	3.2	42	Metallic plates between antennas	Isolation less than 50 dB
[34]	4.4	48	Circularly polarised cross dipole	Isolation less than 50 dB
Prop.	**0.92**	**52.6**	**Cavity backed RMSA at optimised separation**	**High isolation and simple design**

Table 4.
Comparison of Cavity-backed RMSA used in the low-cost GPR with other related works.

buried in semi-dry pebbled soil, an experiment has been carried out for successful detection up to the depth of 65 cm for low-transmitted power (−10 dBm). In general, IEDs are placed at a depth, not more than 50 cm inside the soil, to have a substantial blasting effect. To have the same effect of the explosion with an IED kept at a greater depth, more explosives occupying a larger volume, will be required. Thus, the prototype designed will be effective in detecting IEDs in the field.

The proposed system exhibits highly sensitive performance because of the use of high-isolation and high-gain antennas. High isolation between antennas ensures that the low-reflected power from the small targets is not masked by relatively high-mutual coupling between the transmitting and receiving antennas. **Table 4** compares the cavity-backed antennas used with other related work in yielding high isolation. As evident from **Table 4**, the proposed simple low-cost antenna yields higher isolation than all the refereed antennas. The isolation obtained may be further increased by adding RF absorbent between the antenna.

Most of the reported GPR uses software-based post-processing techniques, such as background subtraction algorithm [9, 10] to alleviate the issue of mutual coupling between antennas. However, the background subtraction algorithm assumes an environment without a target and subtracts it with the environment having the target. Thus, this algorithm is not suitable for real-life scenarios where it cannot be ascertained beforehand about the absence of the target. Considering the on-ground scenarios many works have reported hardware-based pre-processing techniques, such as filtering [15, 16] and time-gating techniques [19, 20]. However, all these techniques make the system complex and cost-prohibitive. In addition, the designed prototype work on a single frequency, unlike the other reported GPR systems [2–8] which operate on a wideband frequency. Hence, it is very simple and low-cost. However, due to a single-frequency operation, it cannot ascertain the depth of the target accurately, neither it can provide a pseudo-image of the target. Remote areas where this prototype is planned to be used are generally devoid of any unnatural foreign substances and thus detection of IEDs even without getting its pseudo-image fulfils the basic requirement. However, to find out the exact depth and pseudo-image of the buried target a band of frequencies would be required so that more information about the target is available. The broad bandwidth will also enable the required resolution in detecting targets. As the objective of the work is to have a simple low-cost device to be used by security forces working in remote areas, only a single operating frequency has been used to design the prototype.

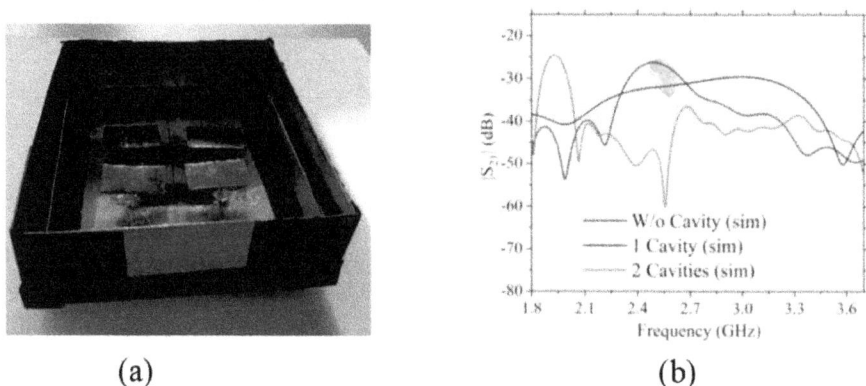

(a)

(b)

Figure 10.
(a) Double-cavity-backed microstrip antenna [36] (b) comparison of isolation (S_{21}) yielded by without cavity-backed, single-cavity-backed, and double cavity-backed stacked multi-resonator microstrip antenna [36].

For broadband continuous-wave operation, the band of frequencies has to be modulated so as to provide a marker for range estimation [1]. Frequency modulated continuous-wave (FMCW) utilising triangular wave [35] is one of the accurate methods for short-range detection. The transmitting and the receiving antennas also have to be designed broadband and high gain with uniform enhanced isolation throughout the band. The concept depicted in **Figure 3(a)** may be extended for providing broadband isolation by introducing multi cavity-backing. Multi-cavity backing has been demonstrated to be versatile in providing a wide BW of 64% for a centre operating frequency of 2.6 GHz, by introducing multi resonances and a high average gain of 12.6 dB by concentrating the RF energy in the desired direction [36]. Multi-cavity backing also demonstrates the potential to yield high uniform isolation of 40 dB [36] by each individual cavity-backing providing the optimum combination of cavity height and separation between the antennas to cause destructive interference as depicted in **Figure 3(a)** for a particular range of narrowband frequencies. The performance of a double cavity-backed antenna has been elaborated in Ref. [36]. **Figure 10(a)** depicts the fabricated double-cavity-backed antenna and **Figure 10(b)** depicts the isolation enhancement by introducing more than one cavity. As evident from **Figure 10(b)**, single-cavity-backing enhances isolation at the lower frequency of operation, while the introduction of one more cavity to make it double-cavity-backed enhances isolation throughout the band of operation.

6. Conclusion

This chapter presents the development and fabrication of a portable compact low-cost CW GPR prototype operating at a single frequency of 920 MHz. A concept of enhancing isolation between co-located antennas has been explained and implemented. The prototype using the enhanced isolation antennas demonstrates the capability to detect both metal and non-metal targets buried in soil as well as in a sandpit. It is sensitive enough to detect a small bunch of wire buried 20 cm in the soil and the maximum depth of detection in a semi-dry soil is 65 cm for a metallic circular plate with a radius of 12.5 cm being used as a target. The prototype is planned to be utilised for detecting IEDs buried in soil for use by security personnel in remote areas.

Author details

Krishnendu Raha* and Kamla Prasan Ray
Defence Institute of Advance Technology, Pune, India

*Address all correspondence to: krishraha@gmail.com

IntechOpen

References

[1] Hary MJ. Ground Penetrating Radar, Theory and Application. Netherland: Elsevier BV; 2008

[2] Iizuka K, Freundorfer AP. Detection of non-metallic buried objects by a step frequency radar. Proceedings of the IEEE. 1983;**71**(02):276-279

[3] Koppenjan SK, Allen CM, Gardner DW, Lee H, Lockwood SJ. Multi-frequency synthetic-aperture imaging with a lightweight ground penetrating radar system. Journal of Applied Geophysics. 2000;**43**:251-258

[4] Lee J, Nguyen C, Scullion T. A novel, compact, low-cost, impulse ground-penetrating radar for non-destructive evaluation of pavements. IEEE Transactions on Instrumentation and Measurement. 2004;**53**(06):1502-1509

[5] Potin D, Duflos E, Vanheeghe P. Landmine ground-penetrating radar signal enhancement by digital filtering. IEEE Transactions on Geoscience and Remote Sensing. 2006;**44**(09):2393-2406

[6] Ye S, Chen J, Liu L, Zhang C, Fang G. A novel compact UWB ground penetrating radar system. In: 14th Int. Conf. Ground Penetrating Radar (GPR). Shanghai, China: IEEE; 2012. pp. 71-75

[7] Cai Z, Lin W, Qi X, Xiao J. Design of low-cost ground penetrating radar receiving circuit based on equivalent sampling. In: IEEE Int. Symp. Circuits and Systems (ISCAS), Sapporo, Japan. 2019. pp. 1-4

[8] Langman A, Dimaio SP, Burns BE, Inggs MR. Development of a low cost SFCW ground penetrating radar. In: International Geoscience and Remote Sensing Symposium. Vol. 4. Lincoln, NE, USA: IEEE; May 1996. pp. 2020-2022

[9] Salvador S, Vecchi G. Experimental tests of microwave breast cancer detection on phantoms. IEEE Transactions on Antennas and Propagation. 2009;**57**(6):1705-1712

[10] Solimene R, Cuccaro A, Dell' Aversano A, Catapano I, Soldovieri F. Background removal methods in GPR prospecting. In: IEEE European Radar Conf., Nuremberg, Germany. 2013. pp. 85-88

[11] Lu B, Song Q, Zhou Z, Wang H. A SFCW radar for through wall imaging and motion detection. In: European Radar Conf. 2011. pp. 325-328

[12] Ahmad F, Amin M. Through-the-wall human motion indication using sparsity-driven change detection. IEEE Transactions on Geoscience and Remote Sensing. 2013;**51**(2):881-890

[13] Verma P, Gaikwad A, Singh D, Nigam M. Analysis of clutter reduction techniques for through wall imaging in UWB range. Progress in Electromagnetics Research B. 2009;**17**:29-48

[14] Gaikwad A, Singh D, Nigam M. Application of clutter reduction techniques for detection of metallic and low dielectric target behind the brick wall by stepped frequency continuous wave radar in ultra-wideband range. IET Radar, Sonar & Navigation. 2011;**5**(4):416-425

[15] Charvat G, Kempel L, Rothwell E, Coleman C, Mokole E. A Through-dielectric radar imaging system. IEEE Transactions on Antennas and Propagation. 2010;**58**(8):2594-2603

[16] Maaref N, Millot P. Array-based ultrawideband through-wall radar: Prediction and assessment of real radar abilities. International Journal of Antennas and Propagation. 2013;**2013**:1-9

[17] Hagness S, Taflove A, Bridges J. Three-dimensional FDTD analysis of a

pulsed microwave confocal system for breast cancer detection: Design of an antenna-array element. IEEE Transactions on Antennas and Propagation. 1999;**47**(5):783-791

[18] Dogaru T, Le C. SAR images of rooms and buildings based on FDTD computer models. IEEE Transactions on Geoscience and Remote Sensing. 2009;**47**(5):1388-1401

[19] Li X, Davis S, Hagness S, Van der Weide D, Van Veen B. Microwave imaging via space-time beamforming: Experimental investigation of tumor detection in multilayer breast phantoms. IEEE Transactions on Microwave the Theory and Techniques. 2004;**52**(8): 1856-1865

[20] Zhao M, Shea J, Hagness S, Van Der Weide D. Calibrated free-space microwave measurements with an ultrawideband reflectometer-antenna system. IEEE Transactions on Microwave and Wireless Components Letters. 2006;**16**(12):675-677

[21] High Frequency VCO Design and Schematics. Available from: http://www.qsl.net/va3iul/High_Frequency_VCO_Design_and_Schematics.htm [Accessed: January 05, 2021]

[22] SGL-0622Z Datasheet. Available from: http://www.alldatasheet.com/datasheet-pdf/pdf/259623/SIRENZA/SGL-0622Z.html [Accessed: January 10, 2021]

[23] Matthaei G, Young L, Jones EMT. Microwave Filters, Impedance Matching Networks, and Coupling Structures. USA: Artech House; 1985. pp. 427-506

[24] AD8347 Datasheet. Available from: http://www.analog.com/static/imported-files/datasheets/AD8347.pdf [Accessed: January 12, 2021]

[25] Kumar G, Ray KP. Broadband Microstrip Antennas. USA: Artech House; 2003. pp. 131-141

[26] Raha K, Ray KP. Designing a cavity backed microstrip antenna with enhanced isolation for the development of a continuous wave ground penetrating radar. Defence Science Journal. 2021;**41**(4):524-534. DOI: 10.14429/dsj.71.15947

[27] Arduino Uno. Available from: https://www.arduino.cc/en/Guide/ArduinoUno [Accessed: January 20, 2021]

[28] Community: LabVIEW Interface for Arduino Setup Procedure. Available from: https://decibel.ni.com/content/docs/DOC-15971 [Accessed: January 22, 2021]

[29] Channabasappa E, Egri R. System and method of using absorber-walls for mutual coupling reduction between microstrip antennas or brick. US patent 7427949B2. 2008

[30] Janssen E, Milosevic D, Herben M, Baltus P. Increasing isolation bet-ween co-located antennas using a spatial notch. IEEE Antennas and Wireless Propagation Letters. 2011;**10**:552-555. DOI: 10.1109/LAWP.2011.2158510

[31] Ghosh CK. A compact 4-channel microstrip MIMO antenna with reduced mutual coupling. International Journal of Electronics and Communications. 2016;**70**(07):873-879. DOI: 10.1016/j.aeue.2016.03.018

[32] Li J, Yang S, Wang C, Joines WT, Lu Q. Metamaterial cavity for the isolation enhancement of closely positioned dual-polarized relay antenna arrays. Microwave and Optical Technology Letters. 2017;**59**(04):857-862. DOI: 10.1002/mop.30413

[33] Tahar Z, Derobert X, Benslama M. An ultra-wideband modified vivaldi antenna applied to through the ground and wall imaging. Progress in Electromagnetics Research C. 2018;**86**:111-122. DOI: 10.2528/pierc18051502

[34] Akbarpour A, Chamaani S. Ultra-wideband circularly polarized antenna for near-field SAR imaging applications. IEEE Transactions on Antennas and Propagation. 2020;**68**(6):4218-4228. DOI: 10.1109/TAP.2020.2975097

[35] Koivumäki P. Triangular and ramp waveforms in target detection with a frequency modulated continuous wave radar [M.S. thesis]. Espoo, Finland: School of Elect. Eng., Aalto Univ.; 2017

[36] Raha K, Ray KP. Broadband high gain and low cross-polarization double cavity-backed stacked microstrip antenna. IEEE Transactions on Antennas and Propagation. IEEE; 10.1109/TAP.2022.3140349